AF375482

BOTANY
BOOK

ISBN 978-1-952704-08-6
Library of Congress LCCN 2026908897

Title: Eileen Chevalier's Typewriter Aalphabet Botany Book / Eileen Chevalier; illustrated by Eileen Chevalier.
Names: Eileen Chevalier, author | Eileen Chevalier, illustrator. | Eileen Chevalier, designer.
Identifiers: Library of Congress 2026908897 | ISBN 978-1-952704-08-6 (hardcover)
Subjects: LCSH Plants—Juvenile literature. |
LCSH: Flowers—Pictoral works—Juvenile literature.
Classification: LCC QK49 .M67 2026 | DDC j52.13—dc23

Edited by Eileen Chevalier
Designed by Eileen Chevalier
The artwork in this book was hand drawn by Eileen Chevalier.
The text is set in Century Gothic, Avenir Next, and Baskerville.

  www.knightkingpress.com

Eileen Chevalier's

TYPEWRITER Aalphabet

BOTANY BOOK

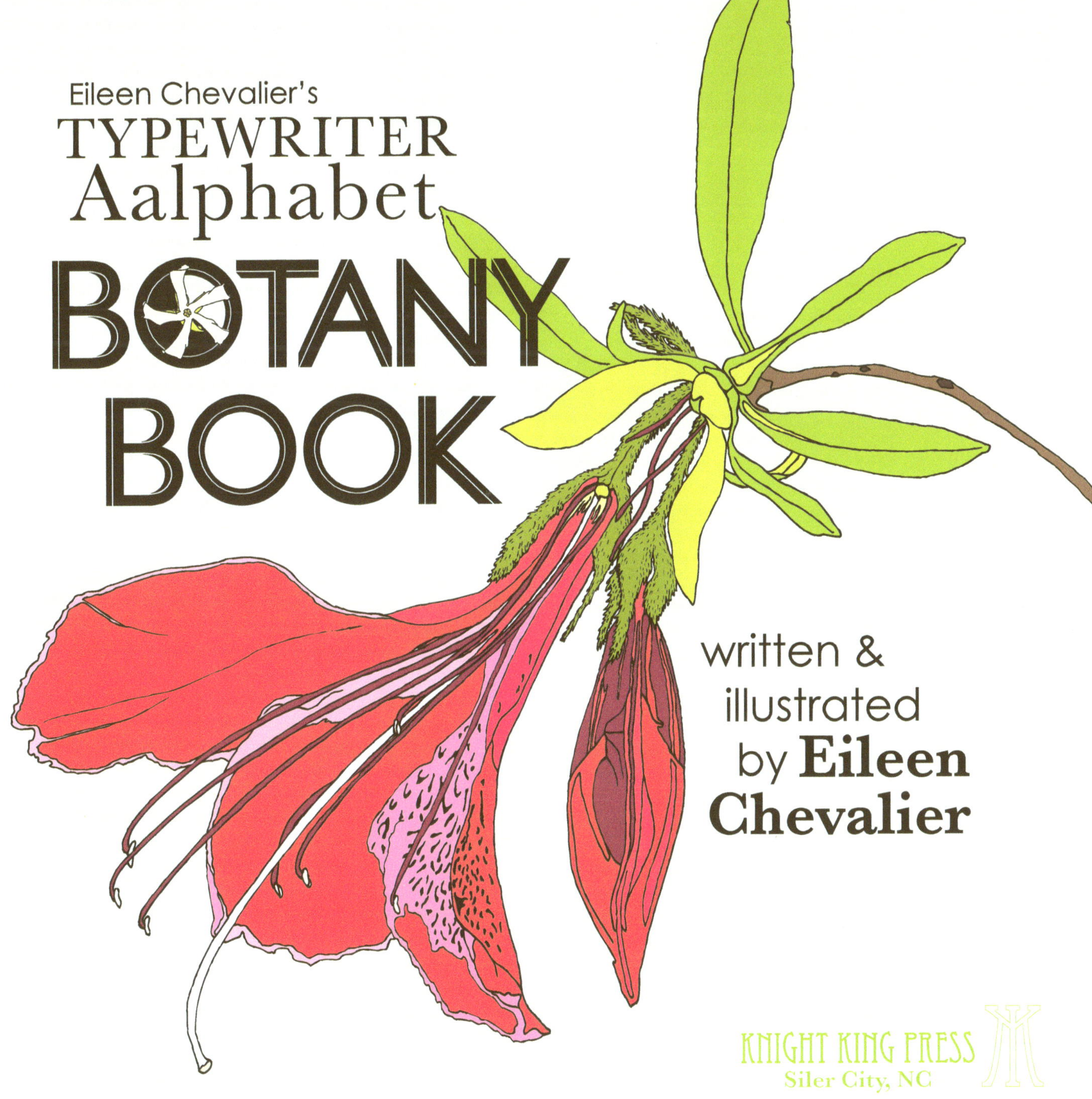

written &
illustrated
by **Eileen Chevalier**

KNIGHT KING PRESS
Siler City, NC

36 HAND-DRAWN, SECTIONED FLOWER ILLUSTRATIONS & BOTANY BASICS

ix petals

7 seven sepals

8
eight bracts

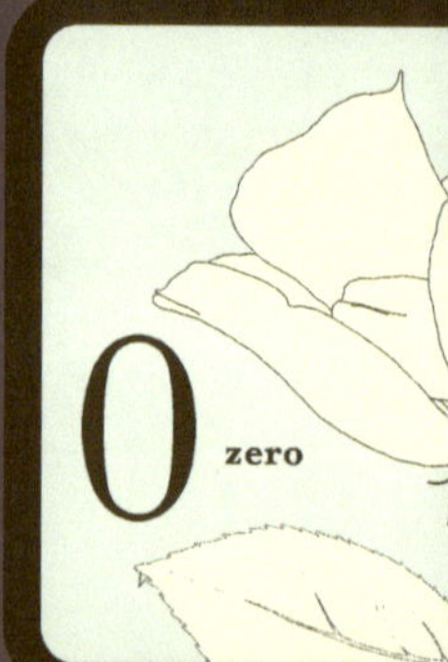
nine nectar guides
9

0 zero

Yy
Yucca

Uu
Ulster Mary

Ii
Iris

Orchid
Oo

g

Hellebore
Hh

Jj
Jasmine

Kudzu
Kk

Lily of the Valley
L

eeding Heart
4 Bb

Nasturtium
Nn

Magnolia
Mm

TABLE OF CONTENTS

Azalea

Azalea

scientific classification
genus: *Rhododendron*
species: *indicum*

Eukaryota - **Domain** - Eukaryota

Azalea

Animalia - **Kingdom** - Plantae

Chordata - **Phylum**.....**Division** - Angiospermae

Mammalia - **Class**.....**Clade** - Eudicotyledoneae

Primates - **Order** - Ericales

Hominidae - **Family** - Ericaceae

Homo - **Genus** - Rhododendron

Human

sapiens - **Species** - indicum

fun fact:
Carl Linnaeus
studied how
to use plants
as medicine
before creating the
Classification System.

Binomial Nomenclature gives the genus and species.

Dicentra formosa is this flower's *scientific name,* but can you tell why its **common name** is bleeding heart?

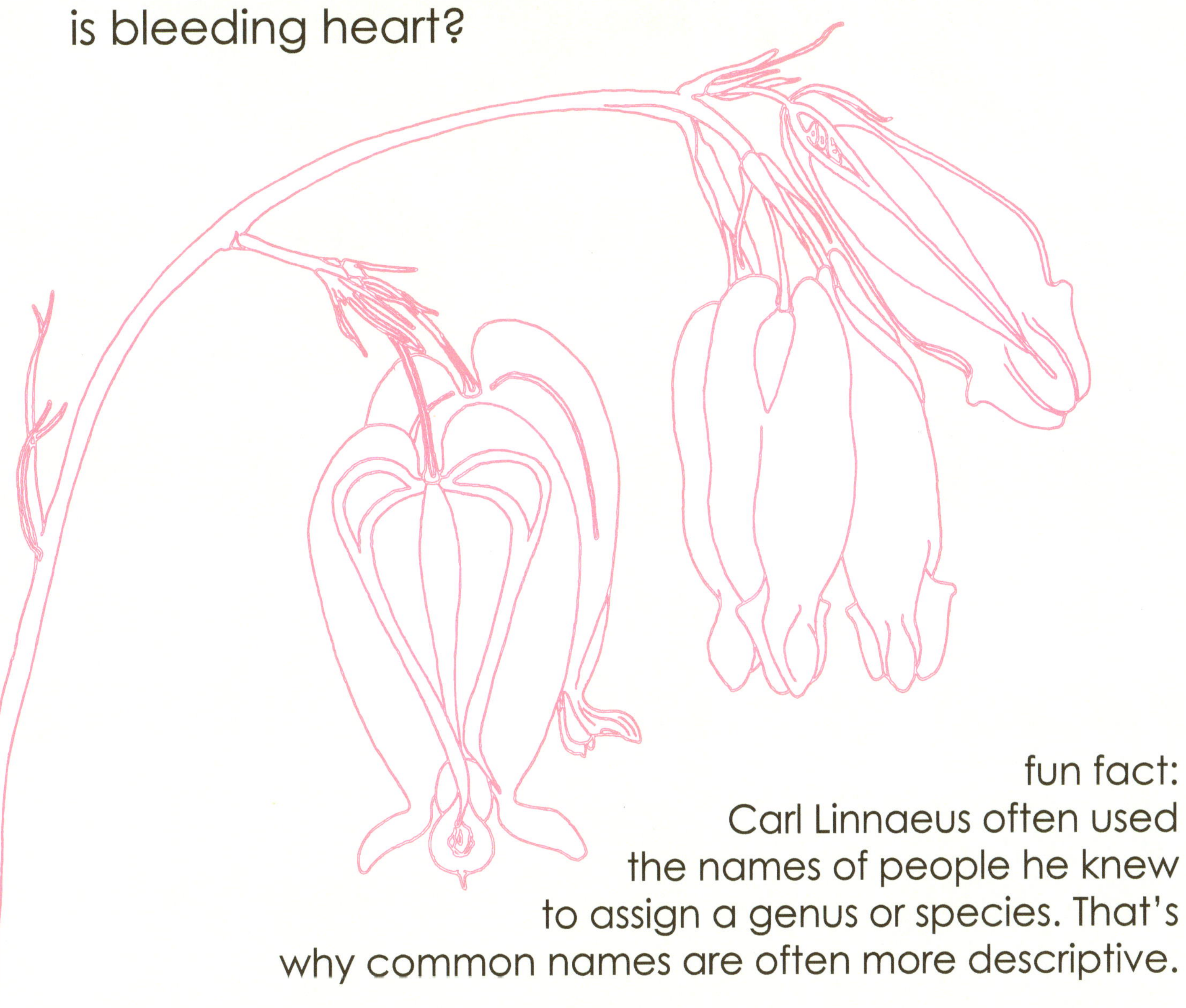

fun fact:
Carl Linnaeus often used
the names of people he knew
to assign a genus or species. That's
why common names are often more descriptive.

Bleeding Heart

Christmas Cactus

Christmas cacti need water to
live and grow, but how do they get it?
How does the floppy stem not topple over?
Plants have **roots** to collect water and nutrients
as well as anchor them in the soil.

fun fact:
Flowers are sometimes named
for when they bloom. Can you
find another flower named
after a time of the year?

Daffodils grow from **bulbs** that live in the ground all year but only come up in the spring.

Bulbs are underground stems attached to special scale leaves that store food for the plant.

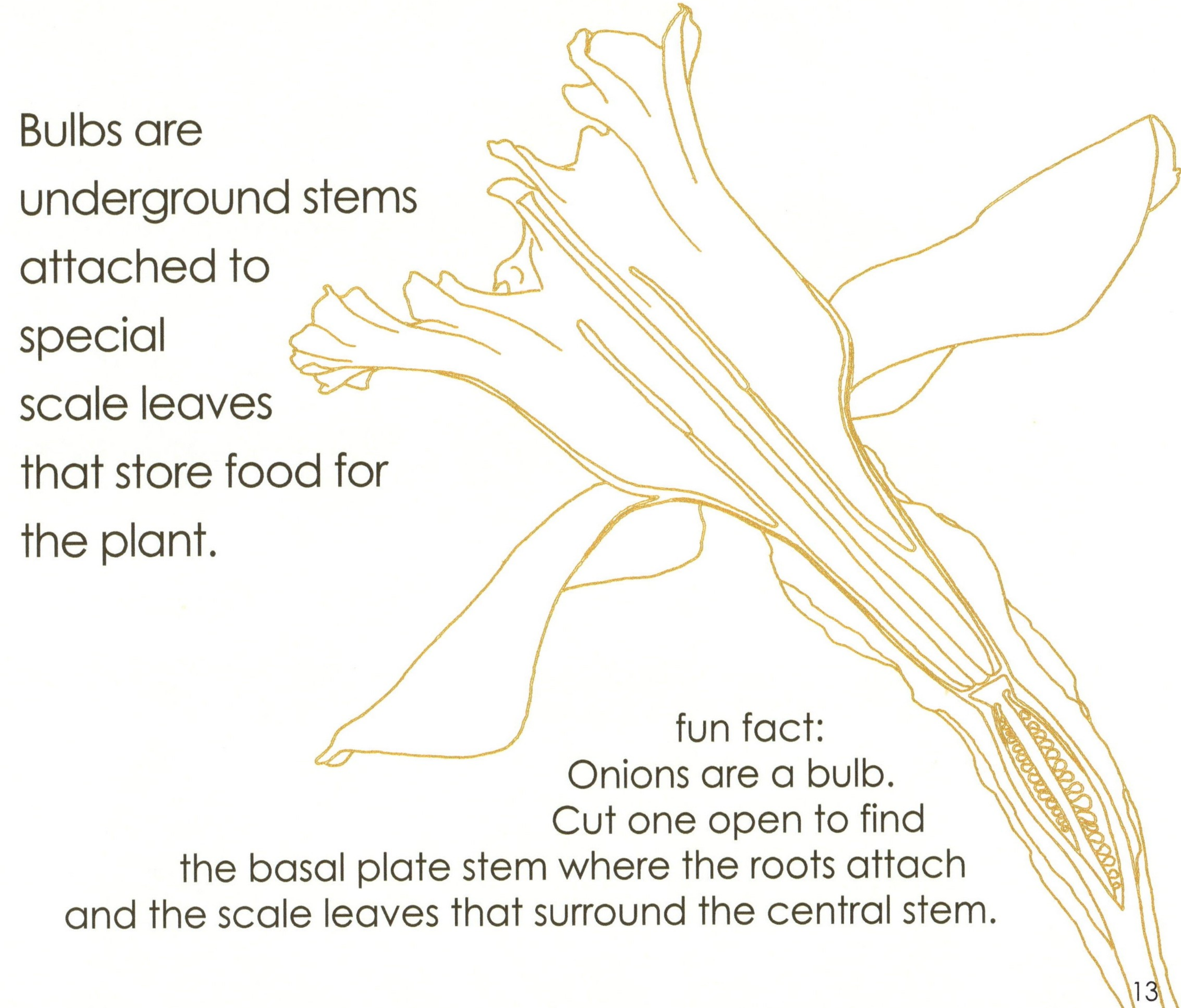

fun fact:
Onions are a bulb.
Cut one open to find
the basal plate stem where the roots attach
and the scale leaves that surround the central stem.

corona corolla
inner tepals in the shape
of a crown (corona)

spathe
special type of bract
(a bract is a special
type of leaf)

scientific classification
genus: Narcissus
species: pseudonarcissus

solitary inflorescence
a single flower on its
peduncle (main stem)

Daffodil

Echinacea

capitulum inflorescence
capitulum = "head"
inflorescence = how flowers are arranged on a stem
Each echinacea flowers produces hundreds of seeds,
each from a single flower on the cone head
pith
spongy tissue
common name
coneflower
scientific classification
genus: Echinacea
species: purpurea

Do you see the purple petals all around the edge of the echinacea? Each one is a tiny flower called a **ray floret**.

Ray florets don't make seeds like the disc florets in the center, but they do attract pollinators.

fun fact:
The way flowers are arranged on a stem is called *inflorescence*. The inflorescence of the echinacea is called *capitulum* or "head" because it contains many tiny flowers (florets) on one stem. Can you find another capitulum inflorescence?

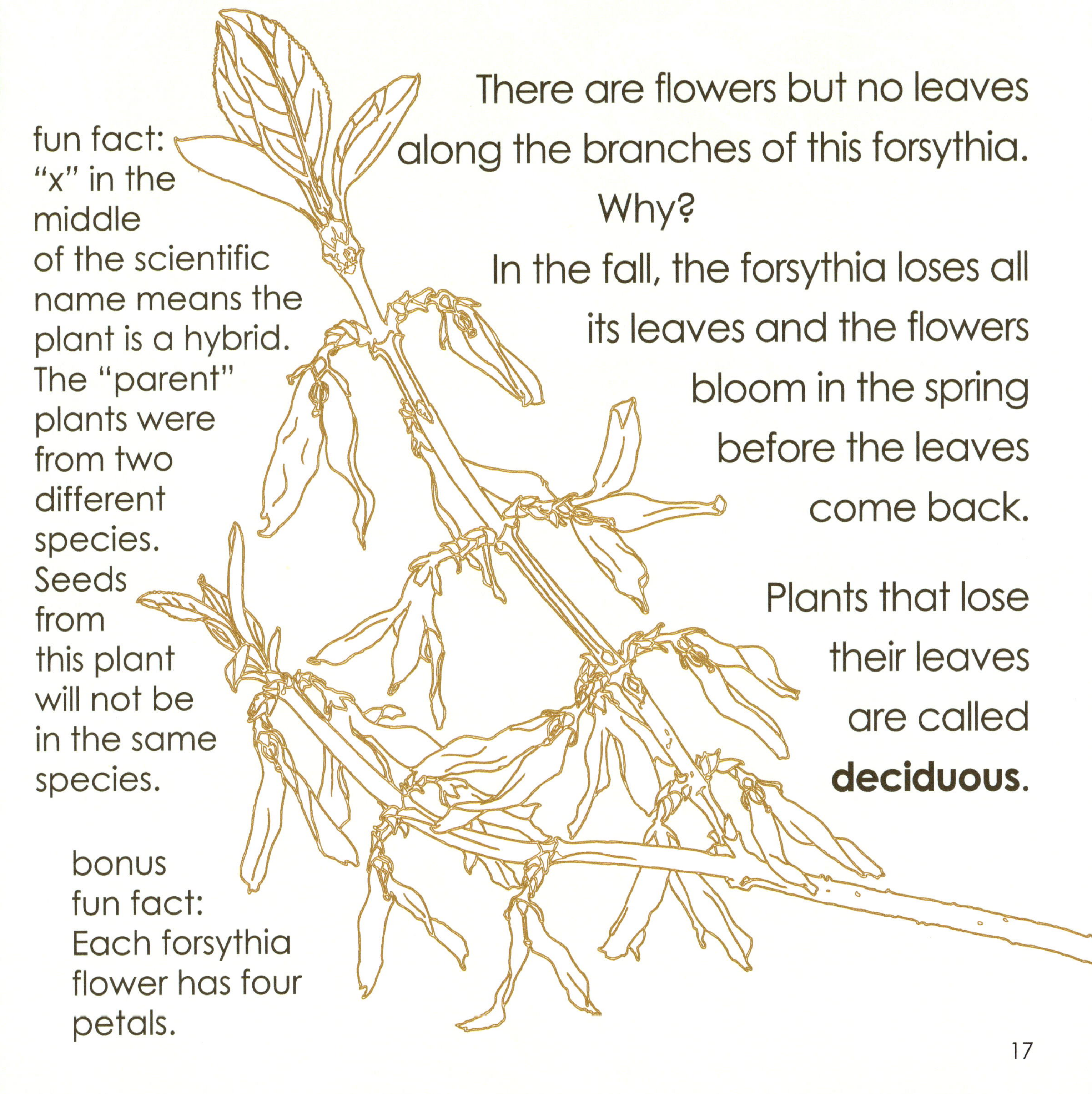

There are flowers but no leaves along the branches of this forsythia. Why?

In the fall, the forsythia loses all its leaves and the flowers bloom in the spring before the leaves come back.

Plants that lose their leaves are called **deciduous**.

Forsythia

scientific classification
genus: *Forsythia*
species: x *intermedia*

spike inflorescence
flowers attached
to a central stem

fun fact
Forsythina was named
after the British botanist
William Forsyth.

18

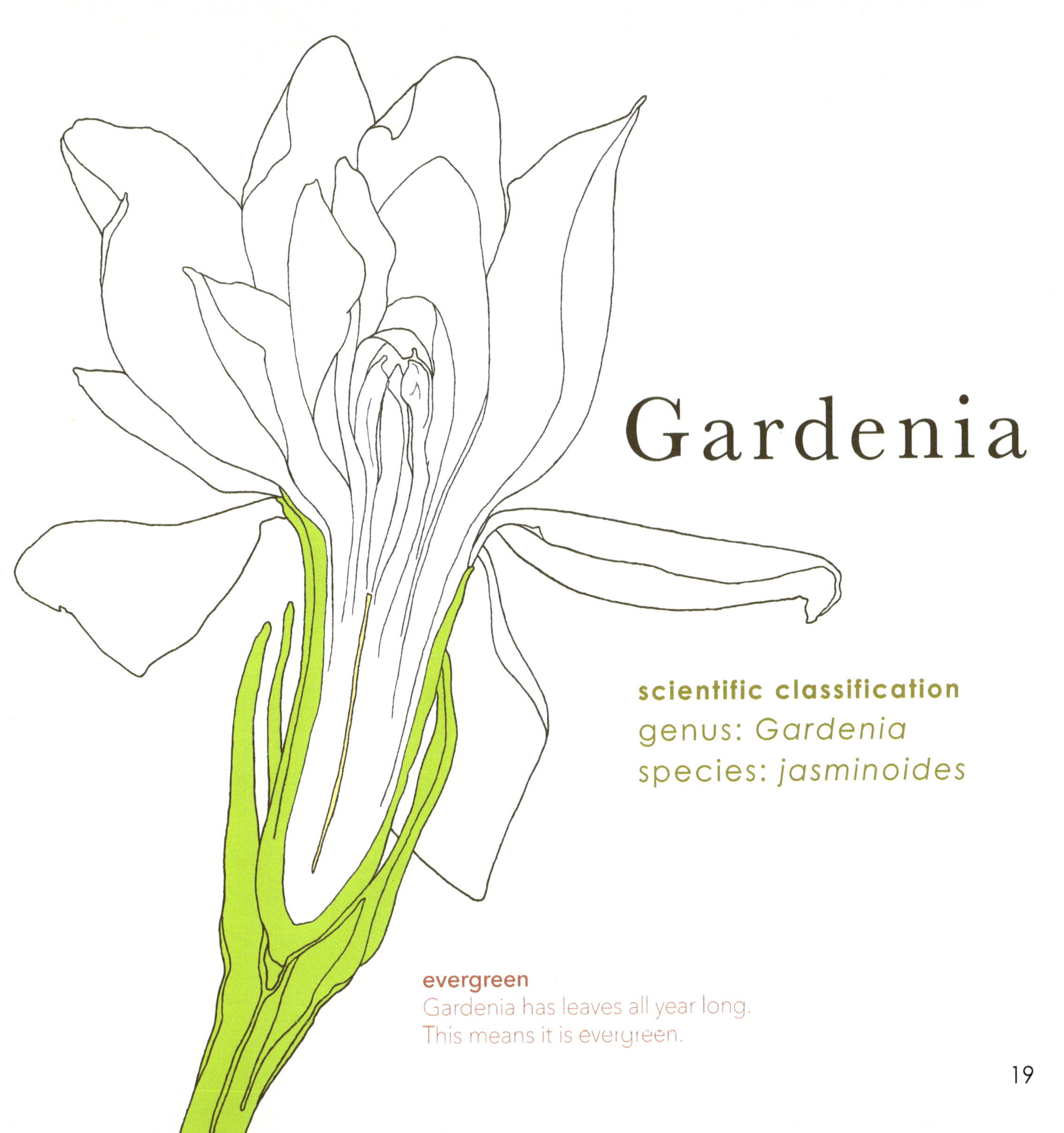

Gardenia

scientific classification
genus: Gardenia
species: jasminoides

evergreen
Gardenia has leaves all year long.
This means it is evergreen.

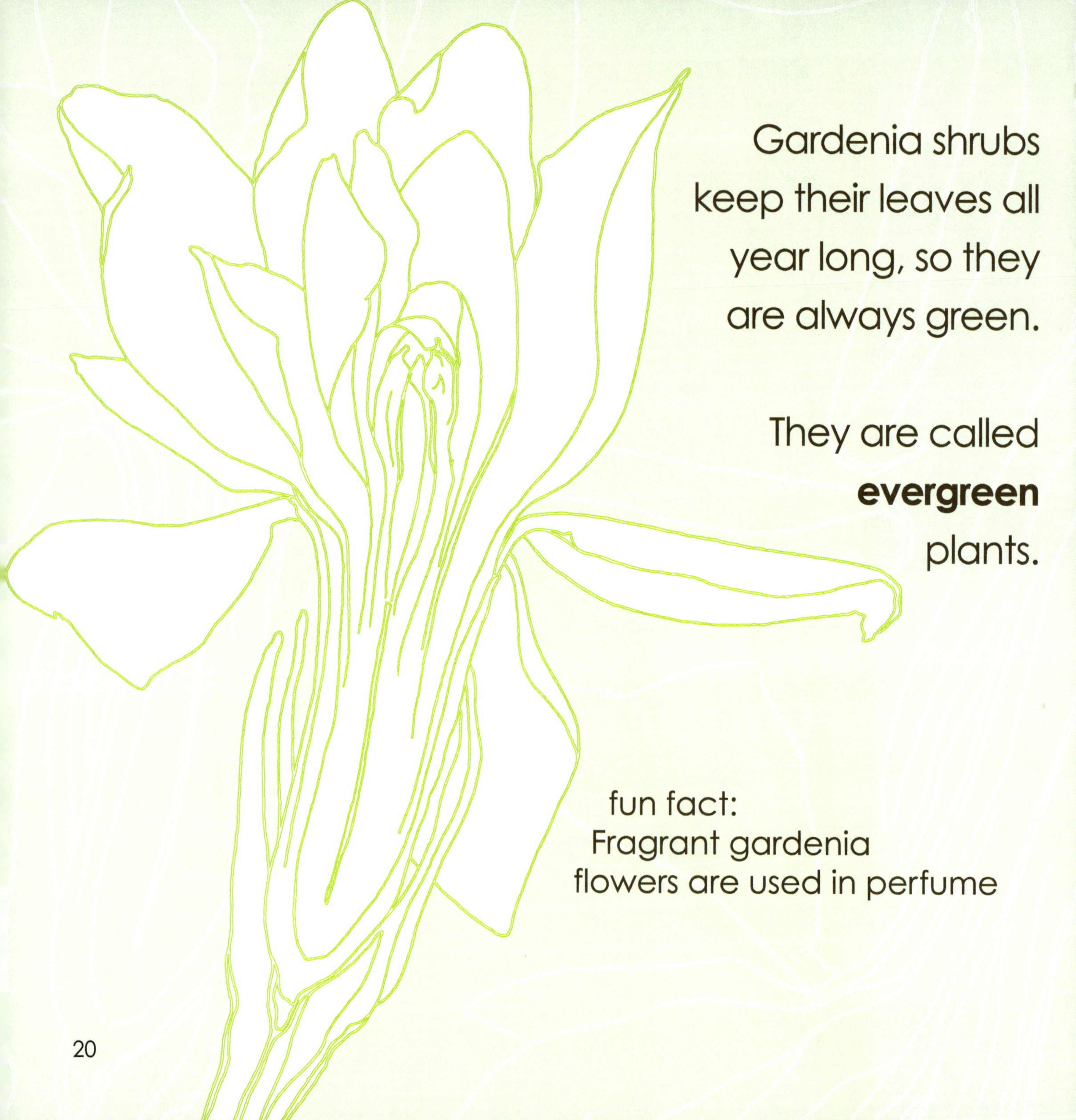

Gardenia shrubs keep their leaves all year long, so they are always green.

They are called **evergreen** plants.

fun fact:
Fragrant gardenia flowers are used in perfume

Do you see how the hellebore flower
is hanging down from the stem?
Some stems, like the gardenia, are **woody** and strong,
but others, like this hellebore
stem, will die each year.
They are called
herbaceous stems.

fun fact:
All the stamens and
nectaries have
already fallen
off this
hellebore.

scientific classification
genus: Helleborus
species: lividus

common name
lenten rose

superior ovary
seeds form at the
top of the flower

sepals
They may look like petals,
but these are sepals.

The tiny petals of the hellebore
also function as nectaries
and fall off when the
flower is polliated.

Hellebore

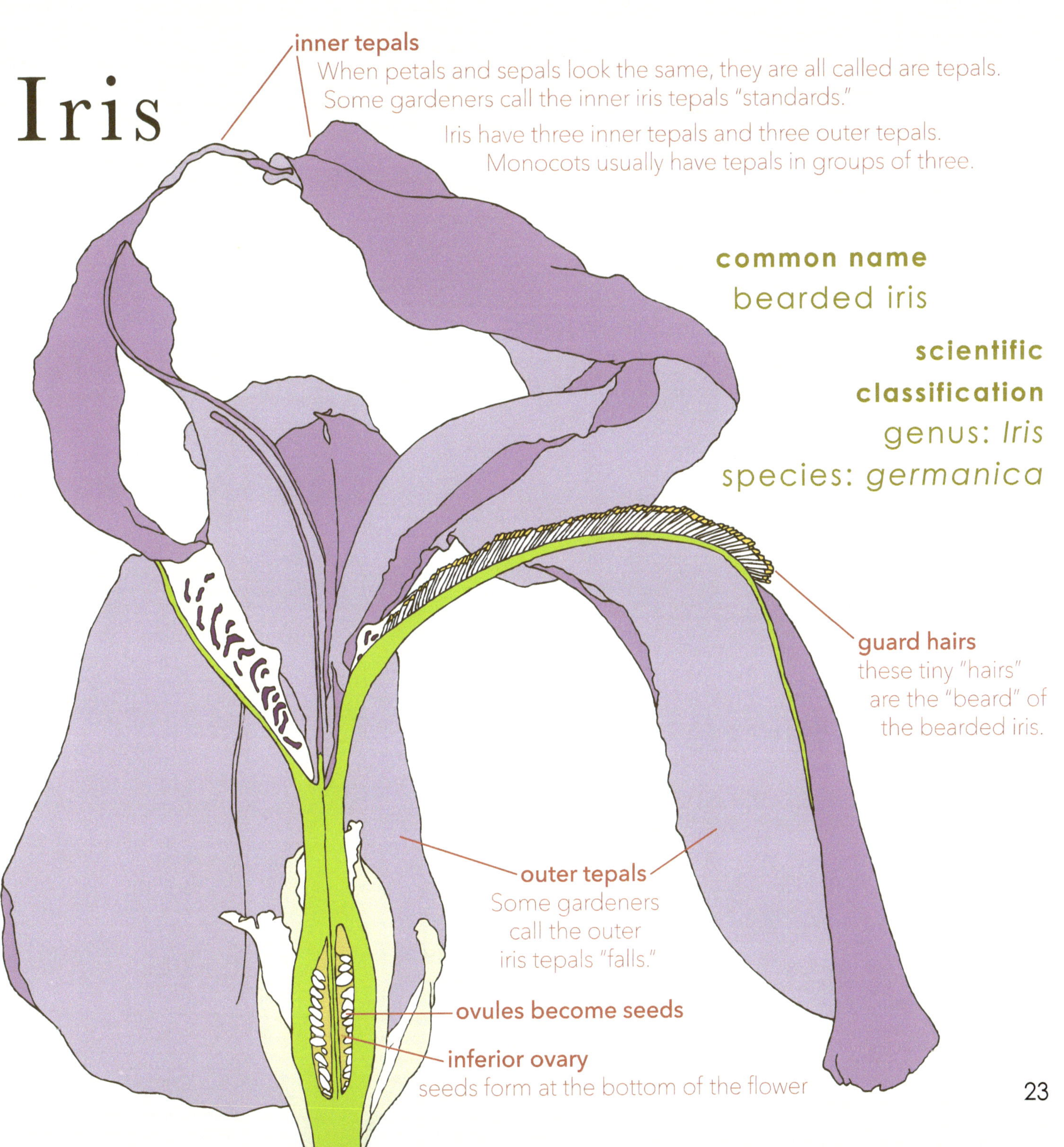

Iris

inner tepals
When petals and sepals look the same, they are all called are tepals.
Some gardeners call the inner iris tepals "standards."

Iris have three inner tepals and three outer tepals.
Monocots usually have tepals in groups of three.

common name
bearded iris

scientific
classification
genus: Iris
species: germanica

guard hairs
these tiny "hairs"
are the "beard" of
the bearded iris.

outer tepals
Some gardeners
call the outer
iris tepals "falls."

ovules become seeds

inferior ovary
seeds form at the bottom of the flower

Iris plants creep along the ground creating new plants from special **stems** called **rhizomes**. Stems have many jobs, including holding up flowers and leaves.

Basal plate stems hold bulbs together and create new plants.

Water & nutrients flow through veins in the stem to leaves and flowers.

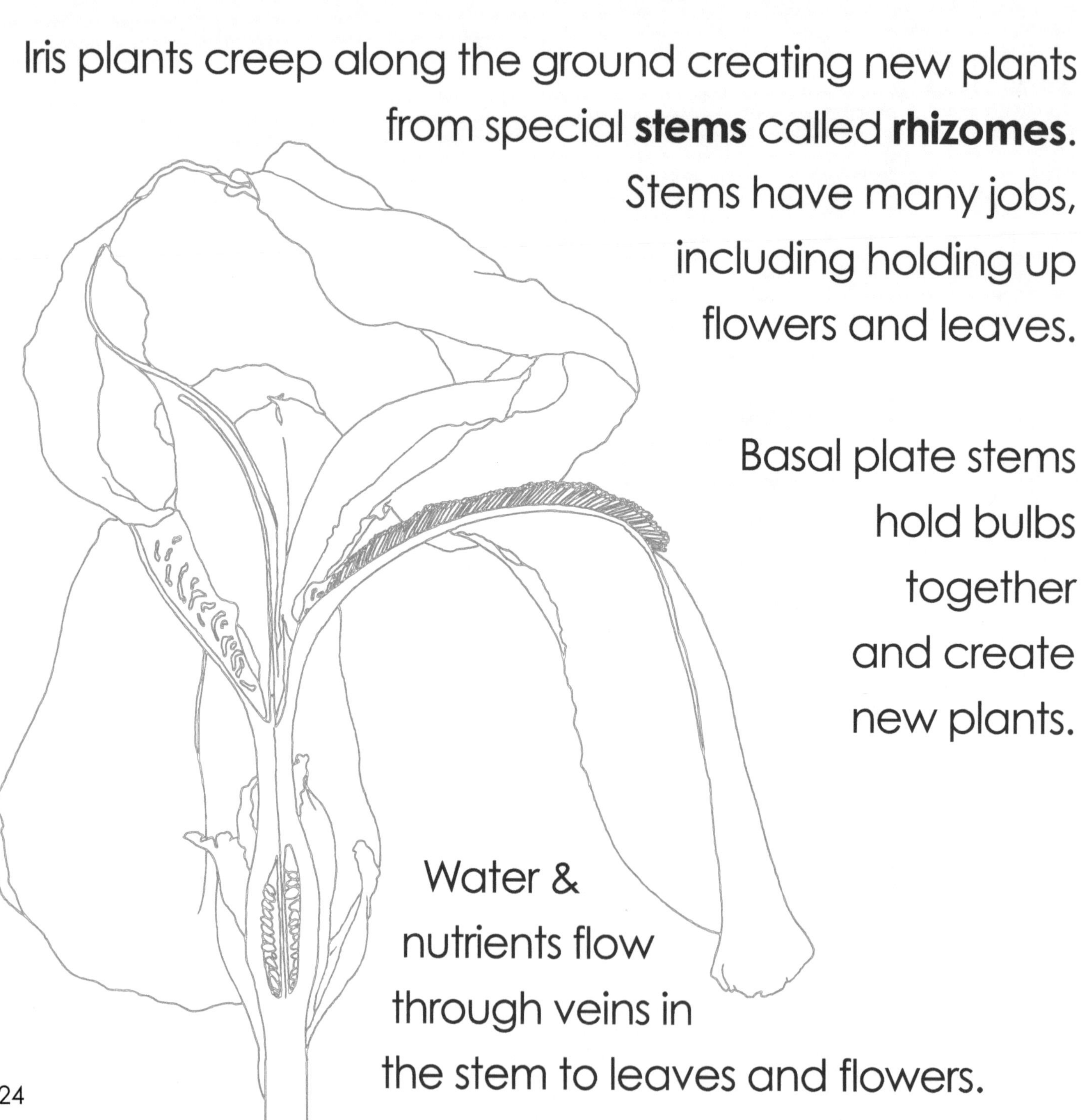

Flowering plants are **angiospermae** because they create covered (angio) seeds (sperm).

Can you see the long "neck" of the flower where the *pistil* is? The pistil is the part of the flower where seeds are formed.

fun fact:
Did you know that the tube inside your neck that allows air to flow to your lungs is called the trachea?

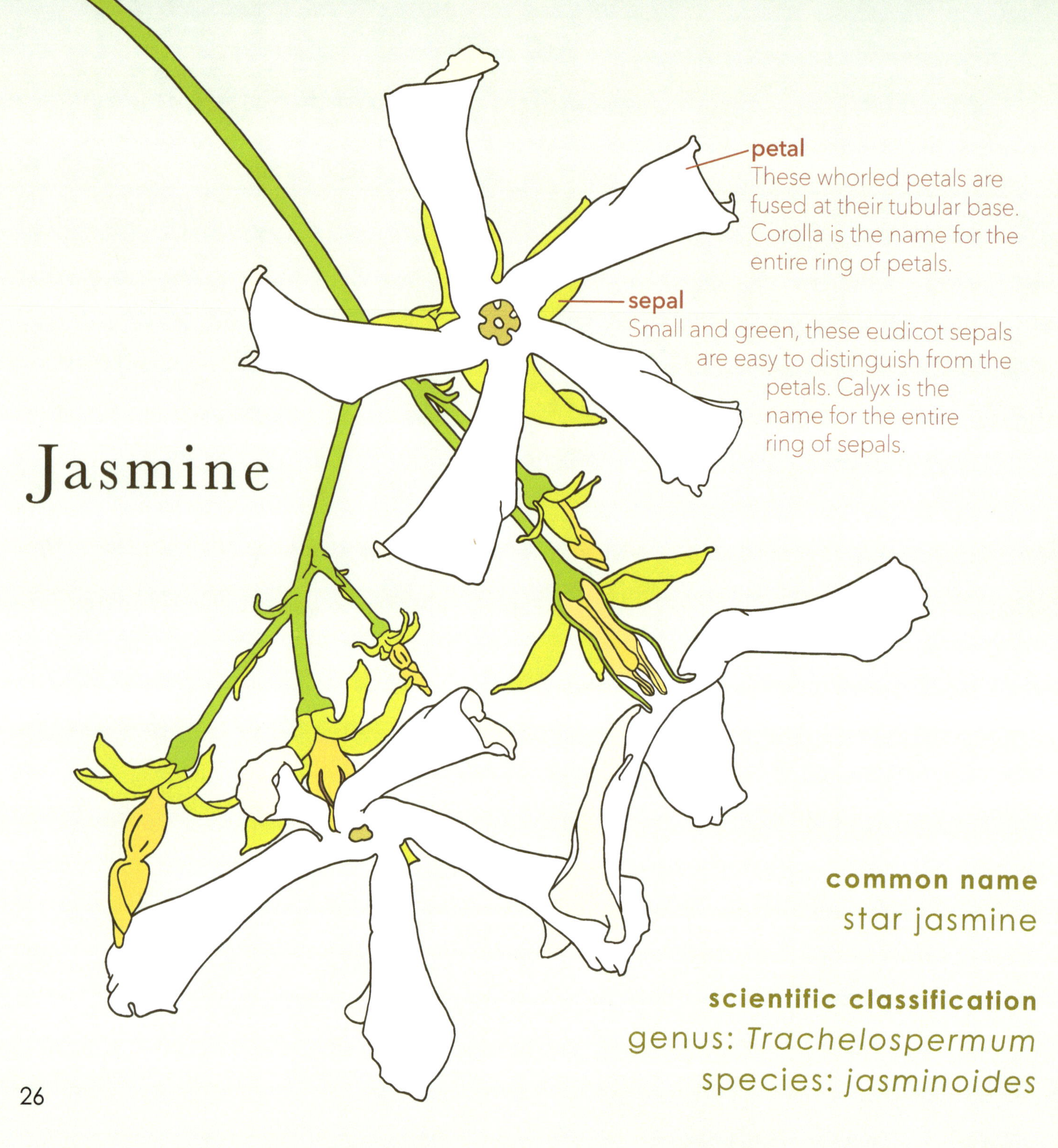

Jasmine
petal
These whorled petals are fused at their tubular base. Corolla is the name for the entire ring of petals.
sepal
Small and green, these eudicot sepals are easy to distinguish from the petals. Calyx is the name for the entire ring of sepals.
common name
star jasmine
scientific classification
genus: Trachelospermum
species: jasminoides
26

perianth
perianth = the calyx + the corolla
calyx = all the sepals of one flower
corolla = all the petals of one flower
Kudzu
scientific classification
genus: Pueraria
species: montana

Do you see the many flowers on this kudzu?

Flowering plants grow from seeds
and then flower to make
new seeds of their own.

This is their **life cycle**.

fun fact:
Kudzu has a raceme
inflorescence, with
many flowers on
a central stem that
bloom from the bottom up.

bonus fun fact:
Kudzu flowers smell like
artificial grape flavoring!

fun fact:
Monocots usually have tepals (instead of petals and sepals)
in groups of three. Lily of the Valley has fused tepals,
so they look like just one, but you can see the
curves of each tepal along
the bottom edge.

How many flowers
are on this stem?

Lily of the Valley is
part of a large group
of angiosperms called
the **Monocotyledoneae clade**.

Mono, from Greek, means *alone* and
a cotyledon is a special tiny leaf that is inside a seed.
When a monocot sprouts, only one leaf appears.

scientific classification
genus: Convallaria
species: majalis
fused tepals
Lilies are monocots, so they usually have tepals in groups of three. Tepals are the name for petals and sepals that are indistinguishable.
Lily of the Valley

tepal
Magnolia is in the Magnoliid clade.
Magnoliids usually have tepals
like the Monocot clade.
Magnolia
scientific classification
genus: Magnolia
species: grandiflora

fun fact:
Magnolias have tepals like a monocot,
a woody stem like a eudicot,
and create seeds
like no other!

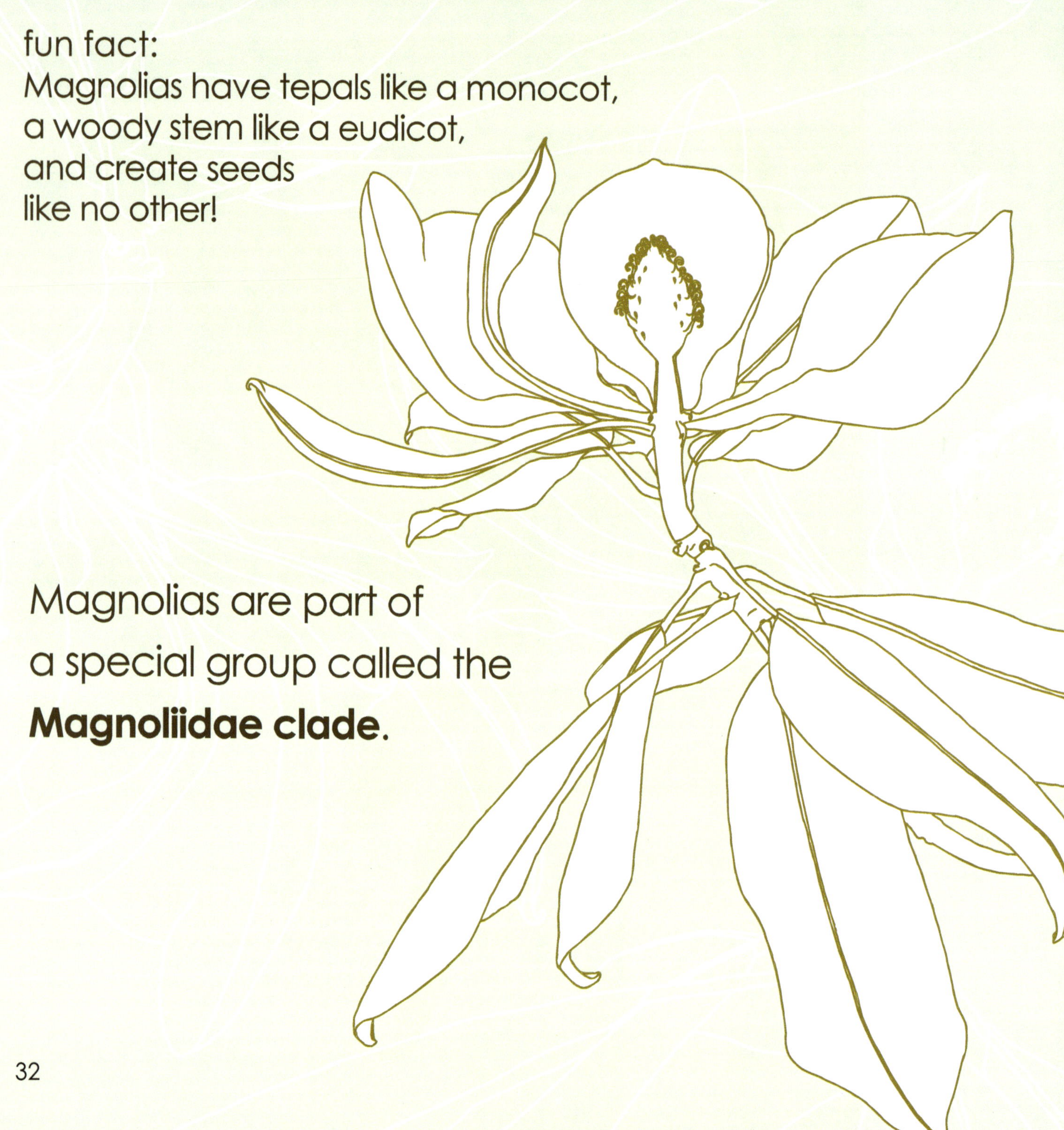

Magnolias are part of
a special group called the
Magnoliidae clade.

Nasturtiums are part of the largest group (clade) of *angiosperms*, the **Eudicotyledoneae Clade**.

"Eudicot" means true (eu-) + two (-di-) seed leaves (cotyledon). Some flowers, such as magnolia, have two cotyledons, but are so different from the eudicots that the old "dicot" clade was divided.

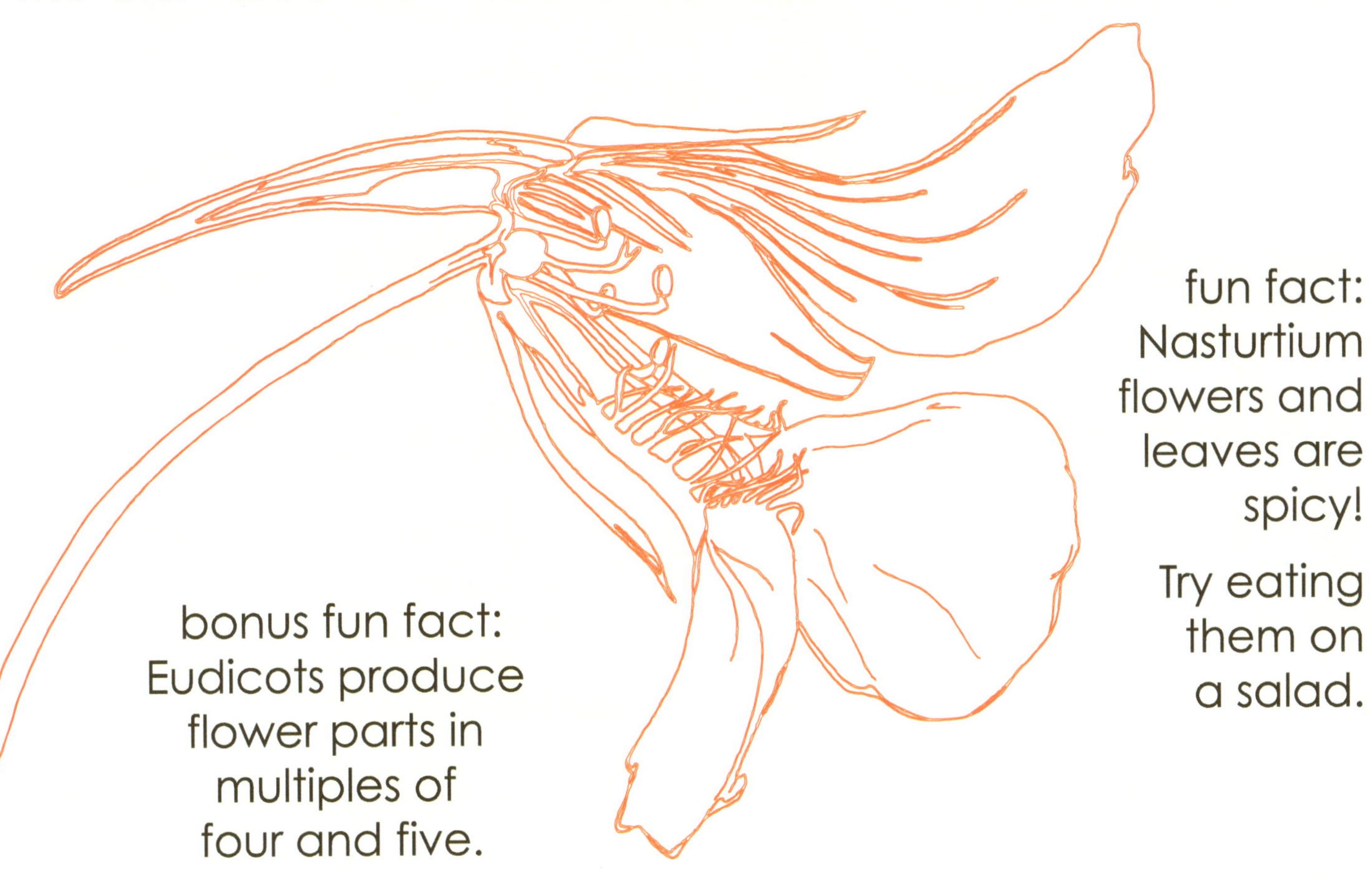

fun fact: Nasturtium flowers and leaves are spicy!

Try eating them on a salad.

bonus fun fact: Eudicots produce flower parts in multiples of four and five.

nectary
Flowers usually
produce nectar to
attract pollinators.

Nasturtium

scientific classification
genus: *Tropaeolum*
species: *majus*

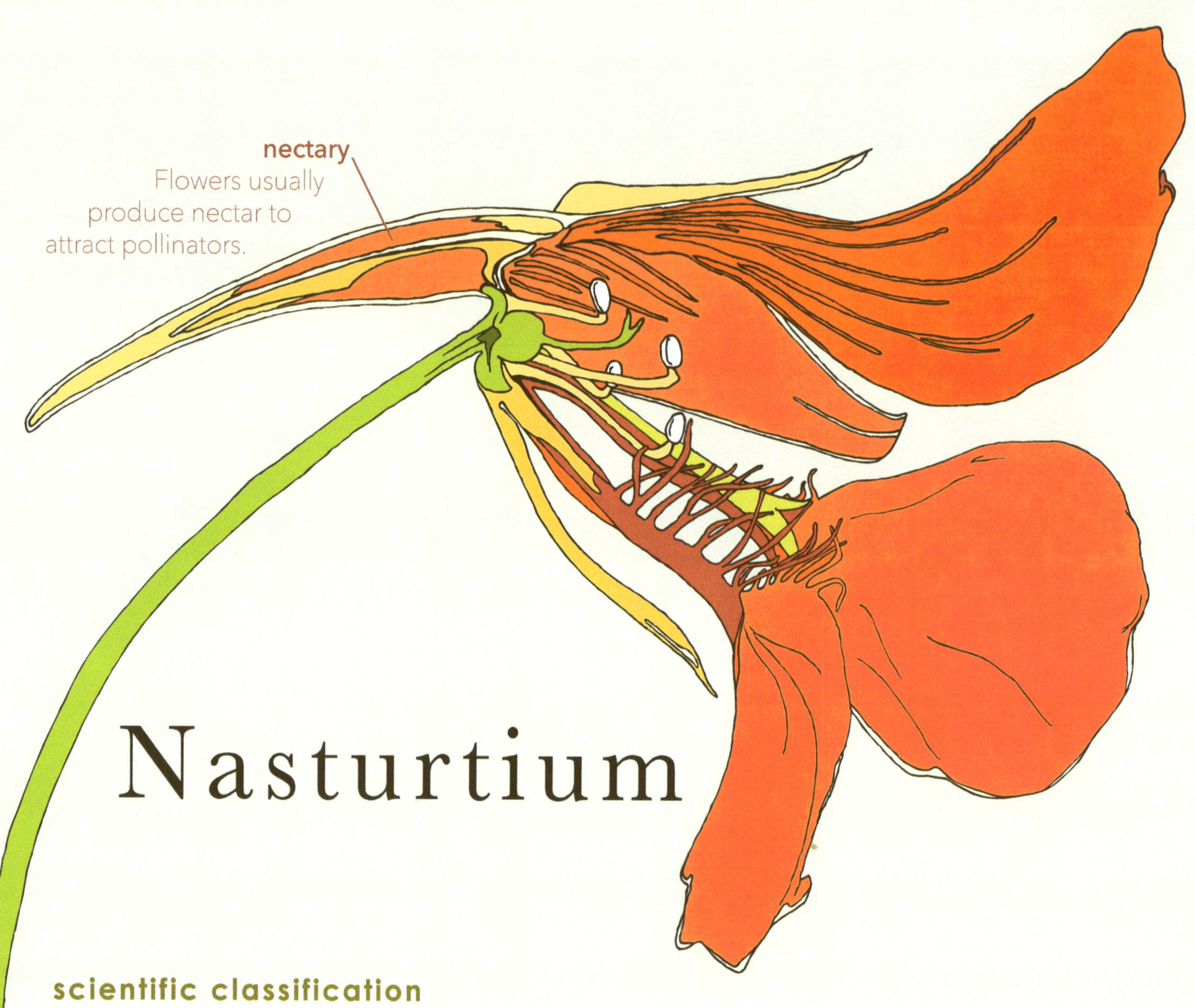

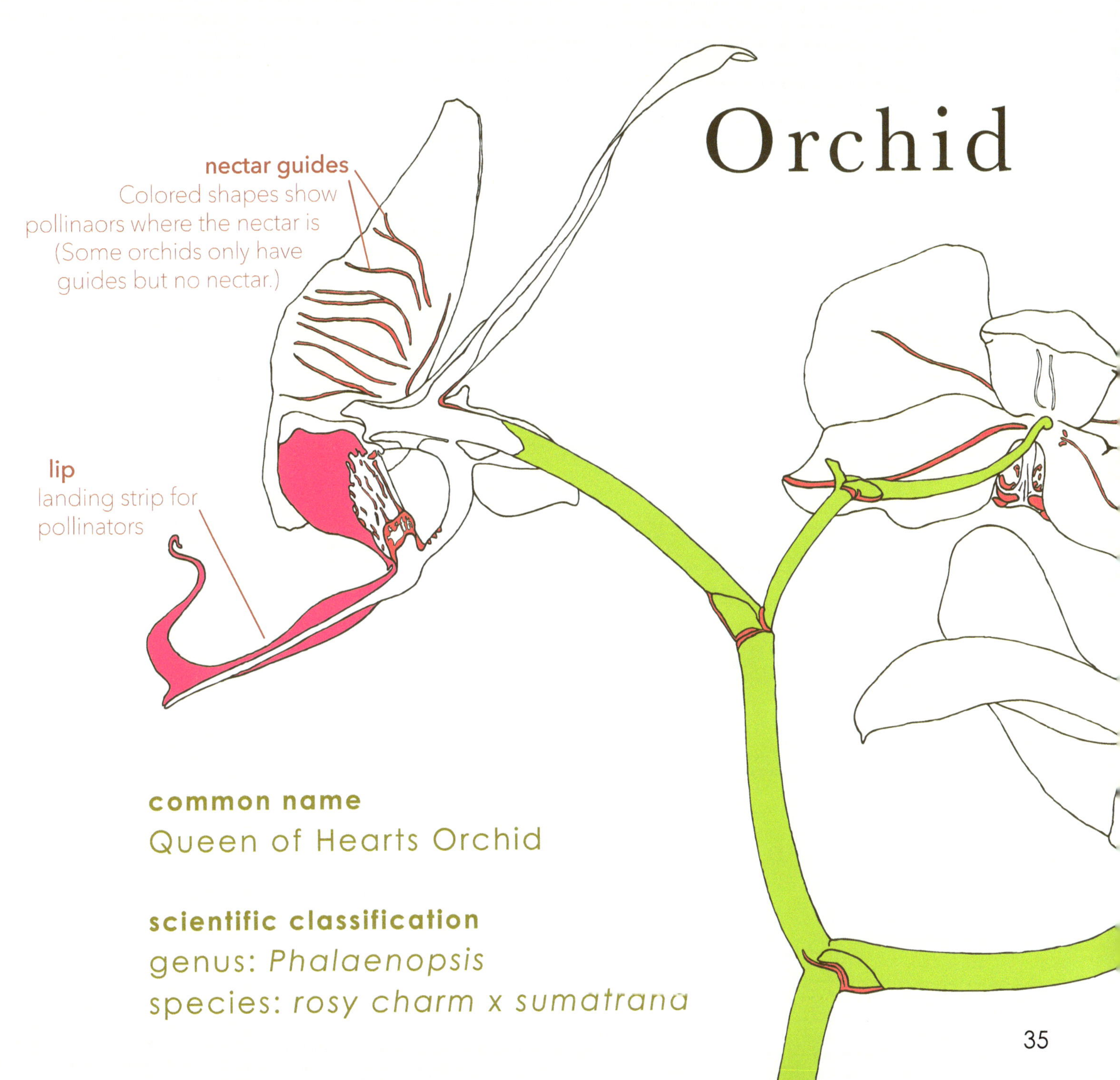

Orchid
nectar guides
Colored shapes show
pollinaors where the nectar is
(Some orchids only have
guides but no nectar.)
lip
landing strip for
pollinators
common name
Queen of Hearts Orchid
scientific classification
genus: Phalaenopsis
species: rosy charm x sumatrana

Flowers come in all shapes and sizes.
Orchid flowers are especially dramatic.

The orchid flower has
bilateral symmetry,
which means
two (bi)
sides (lateral)
have the
same (sym)
measurements (metry).

Have you ever eaten a peach **fruit**?
Every peach was
once a flower!

These peach flowers form
seeds and cover them
with delicious food.
All fruits begin
as flowers.

fun fact:
Some fruits
have many
seeds, but each
peach has only one.

Peach

scientific classification
genus: *Prunus*
species: *persica*

38

fruit
All angiospermae (flowering plants) produce fruit.
Some plants, like the dandelion, produce
dry fruits and others, like this peach,
produce succulent fruit.

scientific classification
genus: *Chaenomeles*
species: *speciosa*

Quince

serrated leaf margin
margin = edge of leaf
branched venation
venation = how veins are arranged in the leaf

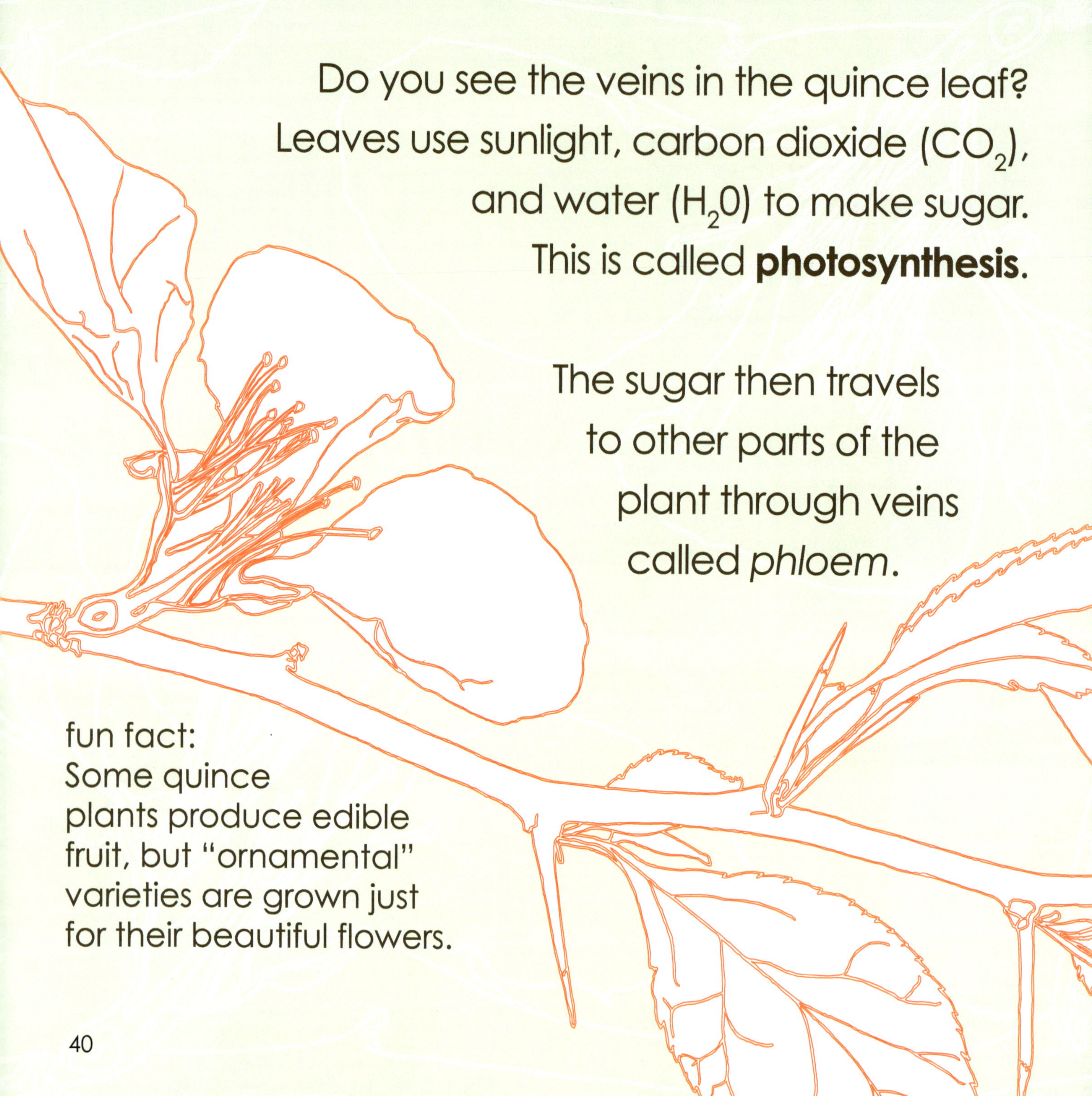

Do you see the veins in the quince leaf?
Leaves use sunlight, carbon dioxide (CO_2),
and water (H_2O) to make sugar.
This is called **photosynthesis**.

The sugar then travels
to other parts of the
plant through veins
called *phloem*.

fun fact:
Some quince
plants produce edible
fruit, but "ornamental"
varieties are grown just
for their beautiful flowers.

Leaves come in all shapes and sizes.

Monocot leaves usually have veins that are parallel, but most eudicot leaves have branching veins like this rose.

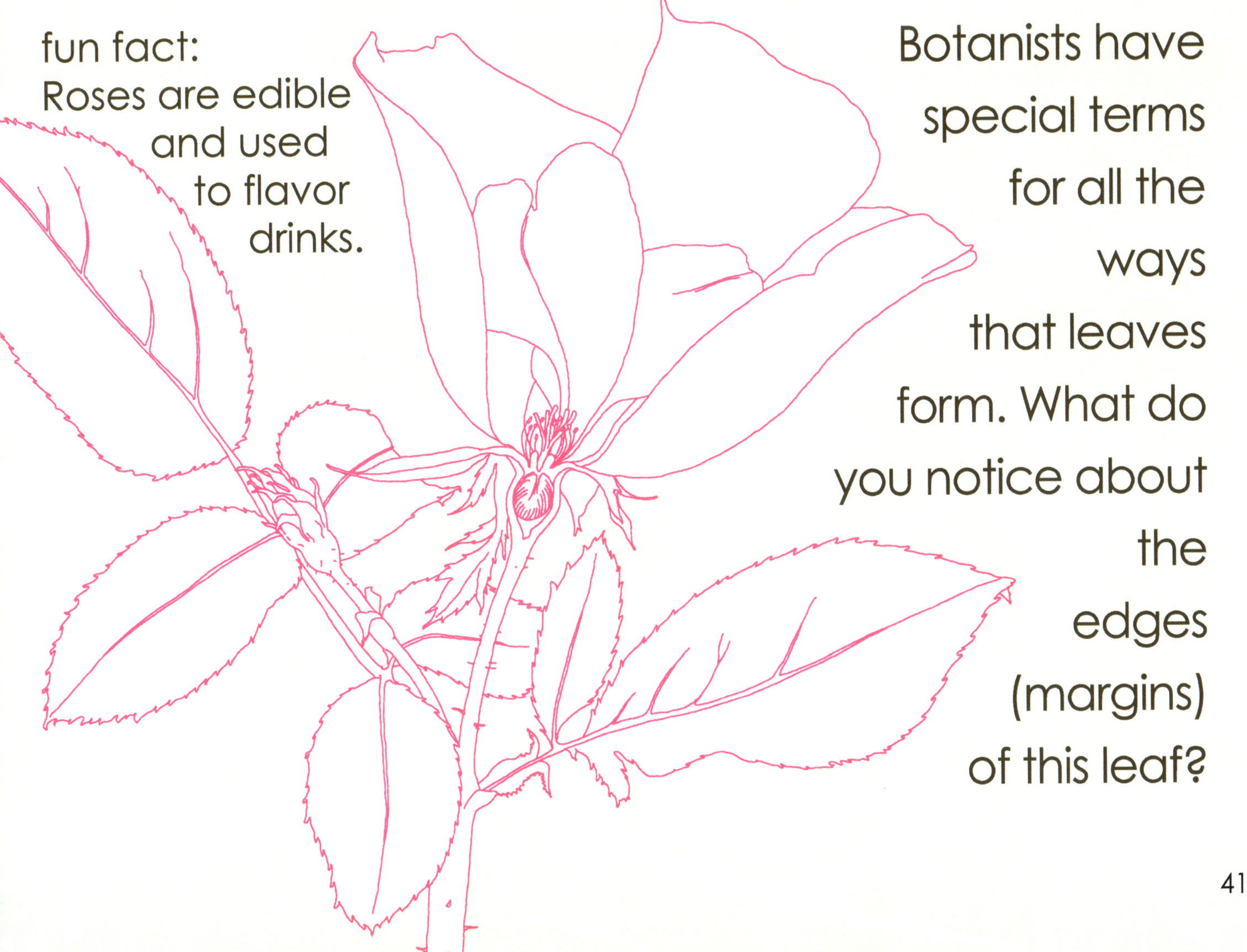

fun fact:
Roses are edible and used to flavor drinks.

Botanists have special terms for all the ways that leaves form. What do you notice about the edges (margins) of this leaf?

prickles
Often mistaken for thorns, prickle emerge all along the stem and are easier to remove.

scientific classification
genus: Helianthus
species: annuus
ray florets
tiny flowers around the outside
of the capitulum inflorescence
that attract pollinators
disc florets
tiny flowers
in the center.
Each disc floret
produces one
sunflower
seed.
Sunflower

Do you see all the tubular
pieces attached to the
center of the sunflower?

Each one is a tiny flower
called a **disc floret**
that will create a
single sunflower seed.

fun fact:
Flowers have petals
in multiples of 3, 4, or 5,
never just 1 or 2. These floret
petals have fused together so they
look like just one. Tricky!

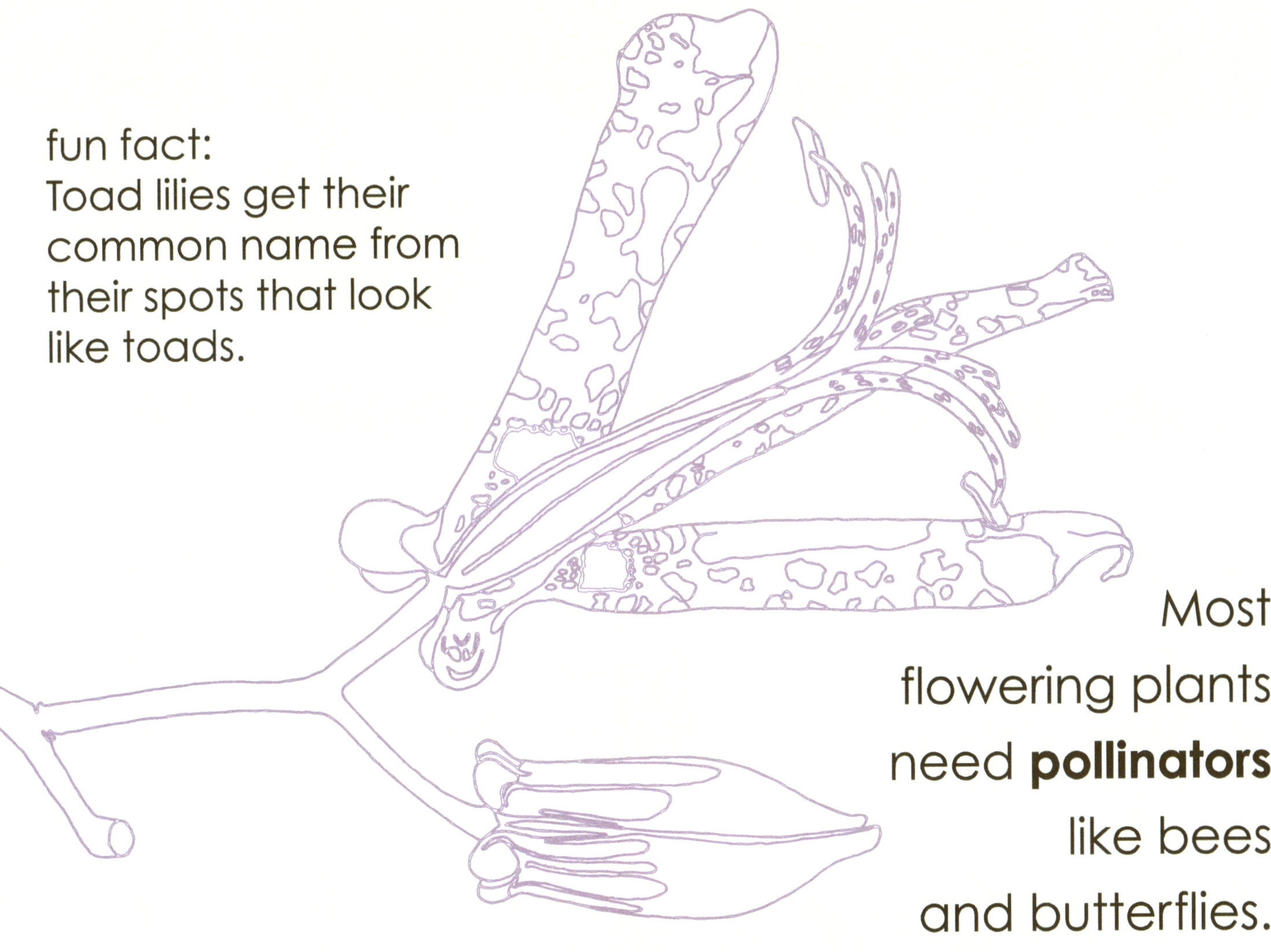

fun fact:
Toad lilies get their
common name from
their spots that look
like toads.

Most
flowering plants
need **pollinators**
like bees
and butterflies.
Pollinators move pollen from the
anther of the stamen to the stigma of the pistil.
This is how the flowers make seeds!

scientific classification
genus: *Tricyrtis*
species: *hirta*

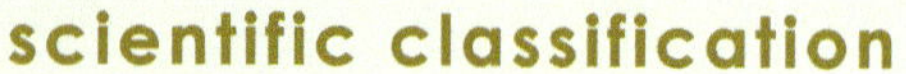

Toad Lily

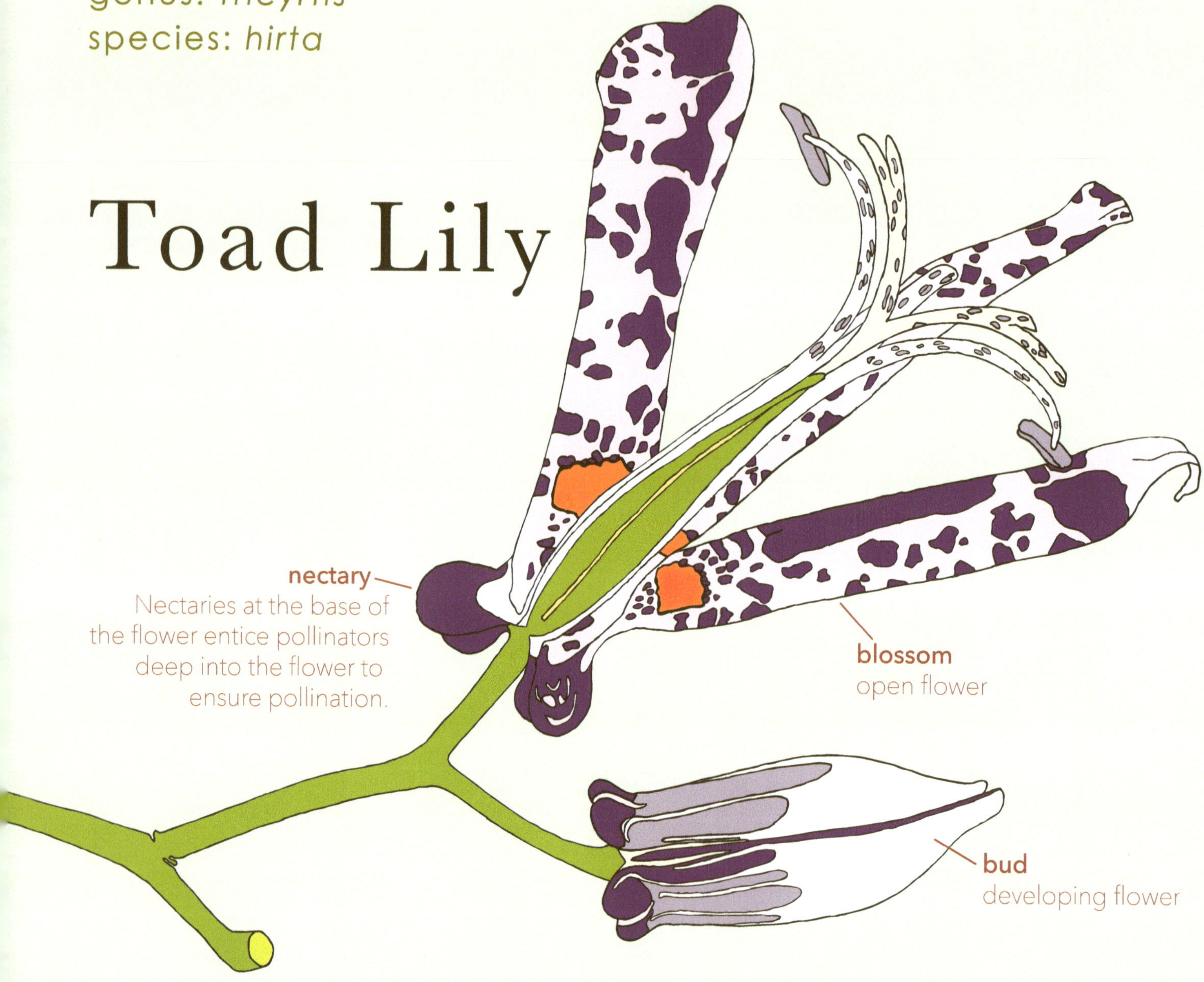
nectary
Nectaries at the base of the flower entice pollinators deep into the flower to ensure pollination.
blossom
open flower
bud
developing flower

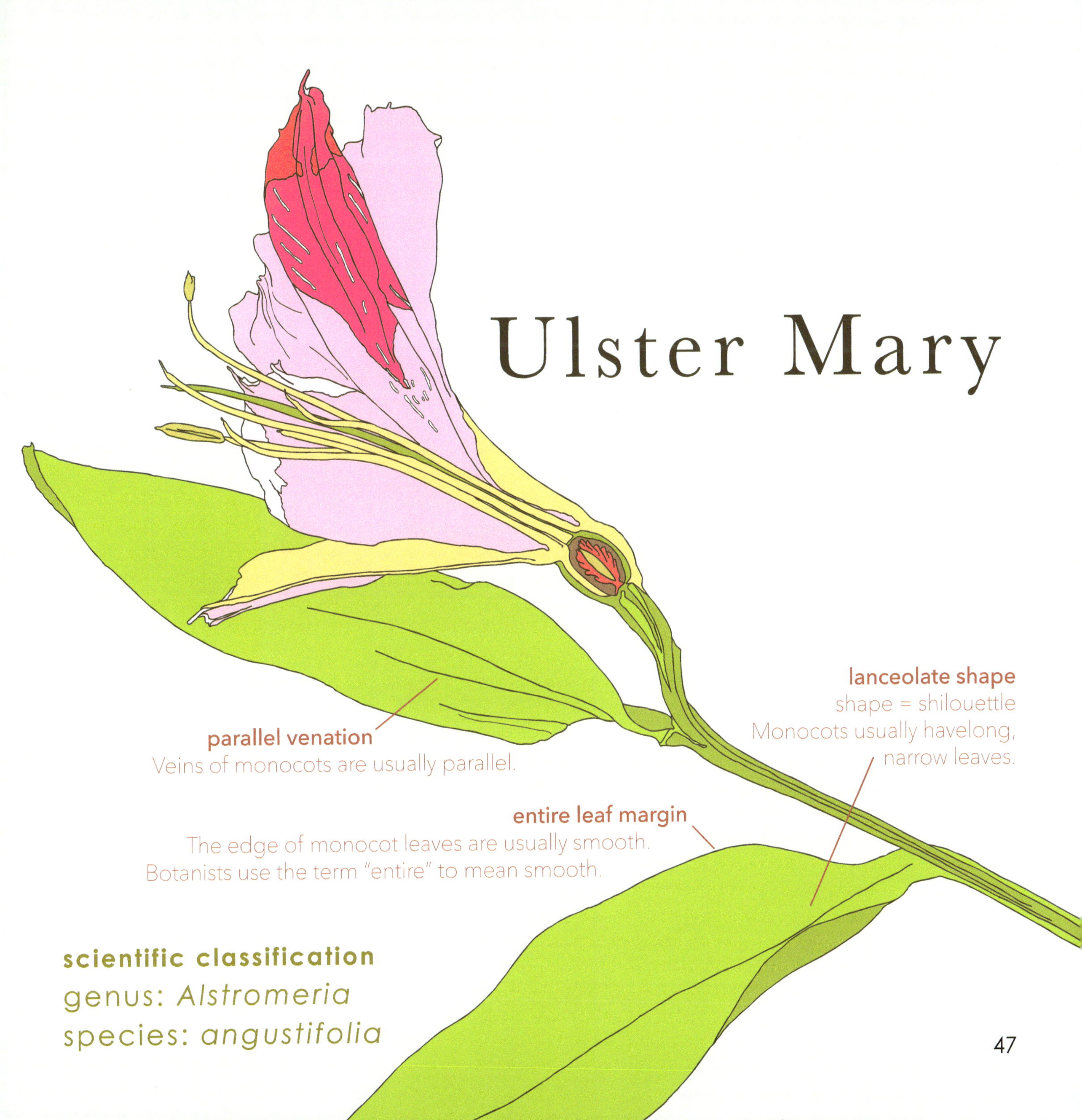

Ulster Mary
lanceolate shape
shape = shilouettle
Monocots usually havelong, narrow leaves.
parallel venation
Veins of monocots are usually parallel.
entire leaf margin
The edge of monocot leaves are usually smooth.
Botanists use the term "entire" to mean smooth.
scientific classification
genus: Alstromeria
species: angustifolia
47

Plants make seeds so their seeds can make new plants. This is called **sexual reproduction**.

Do you see the round ovary at the base of the tepals (petal/sepal) in this Ulster Mary flower? That's where the pollen that travelled down the style of the pistil meets with an ovule to create a new seed.

fun fact:
The Alstromeria was named after botanist Baron Claus von Alstomer in 1753.

Violets are tiny flowers that grow close to the ground. As well as making seeds with flowers, many angiosperms can use **vegetative reproduction** to make new plants.

Vegetative reproduction uses roots, stems, and leaves to create new plants that are identical to the parent plant.

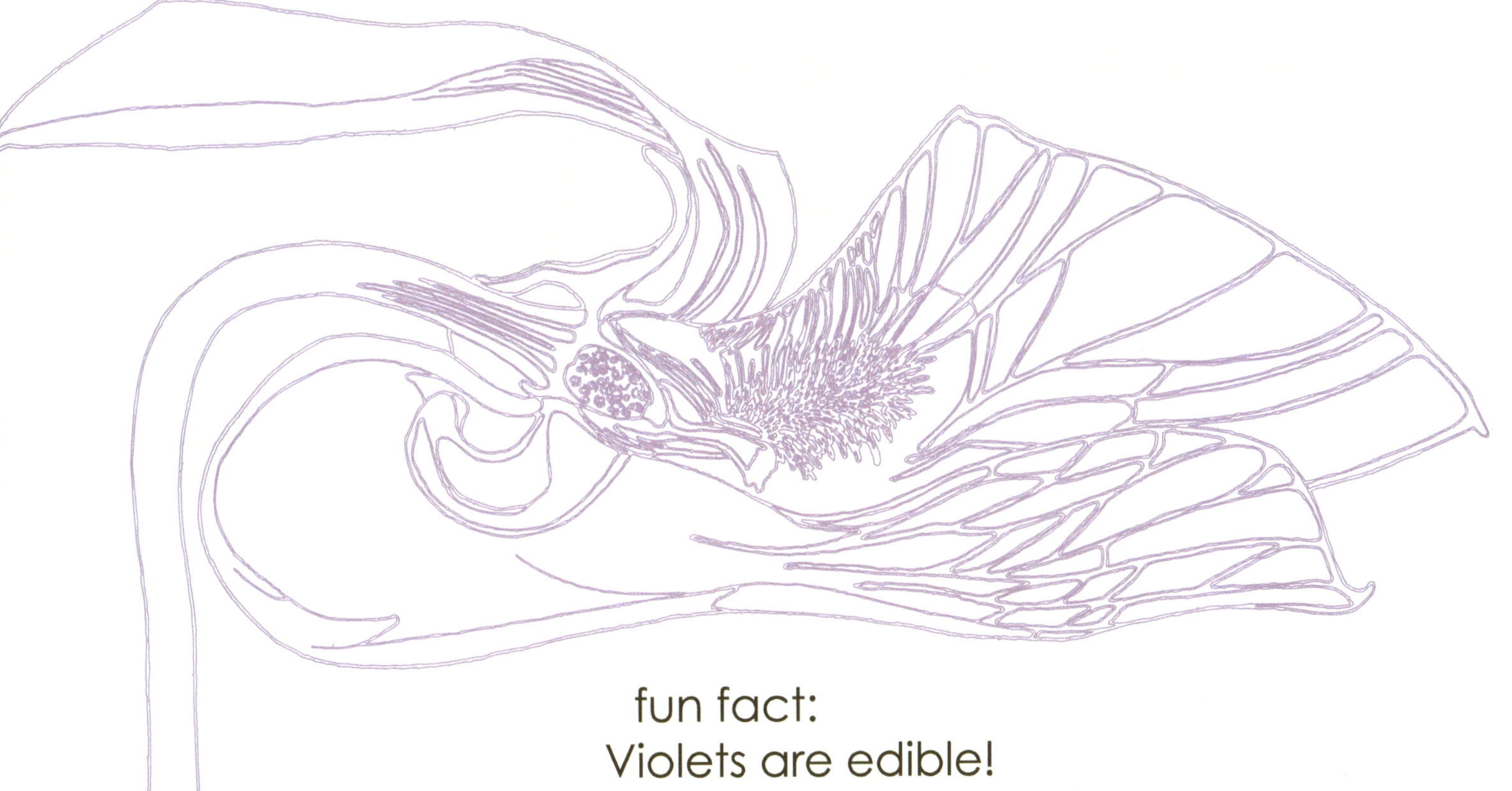

fun fact:
Violets are edible!

Violet

herbaceous perennial
dies back to the roots each year

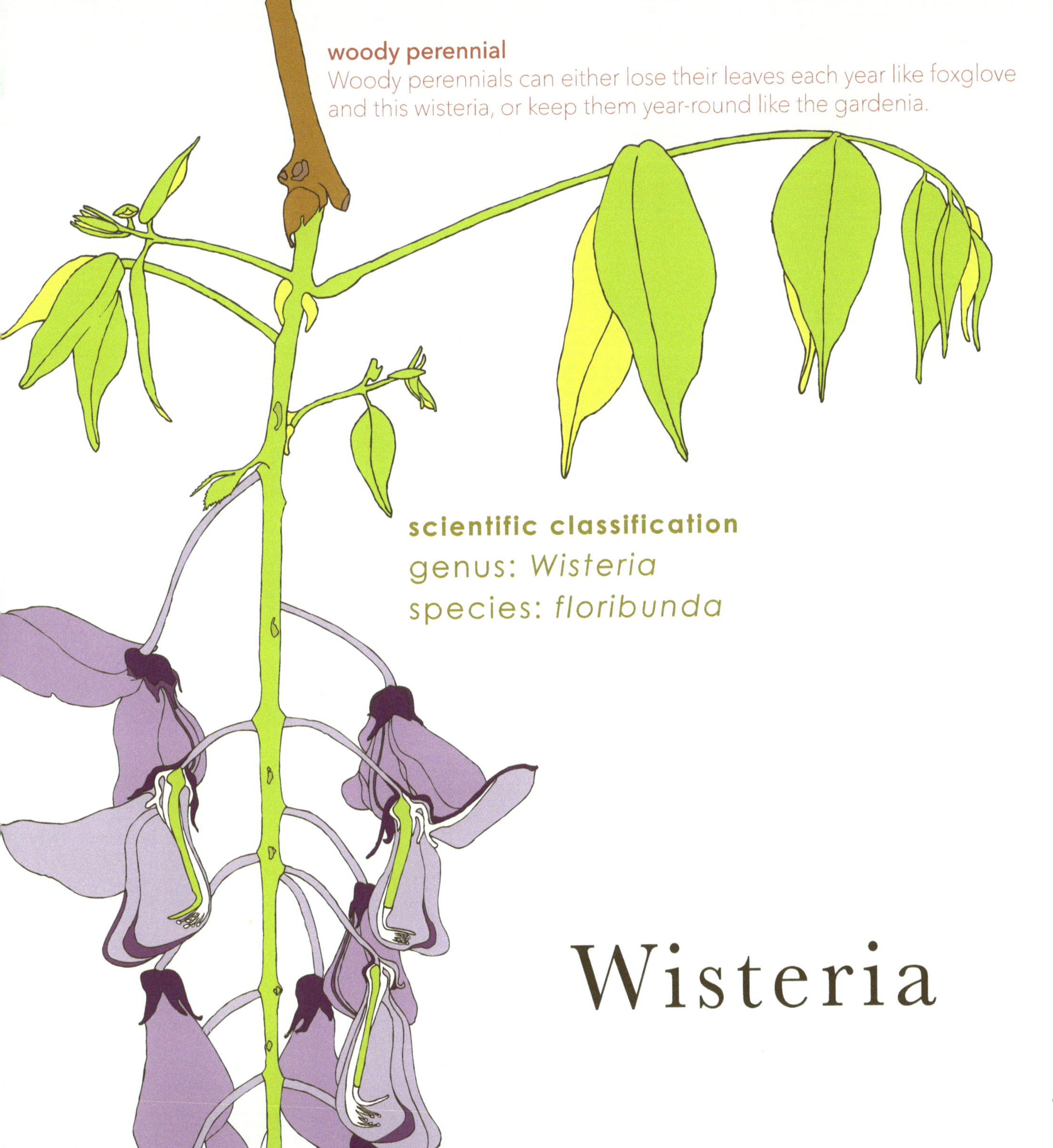

woody perennial
Woody perennials can either lose their leaves each year like foxglove
and this wisteria, or keep them year-round like the gardenia.

scientific classification
genus: Wisteria
species: floribunda

Wisteria

6
six petals

7
seven sepals

8
eight bracts

nine nectar guides
9

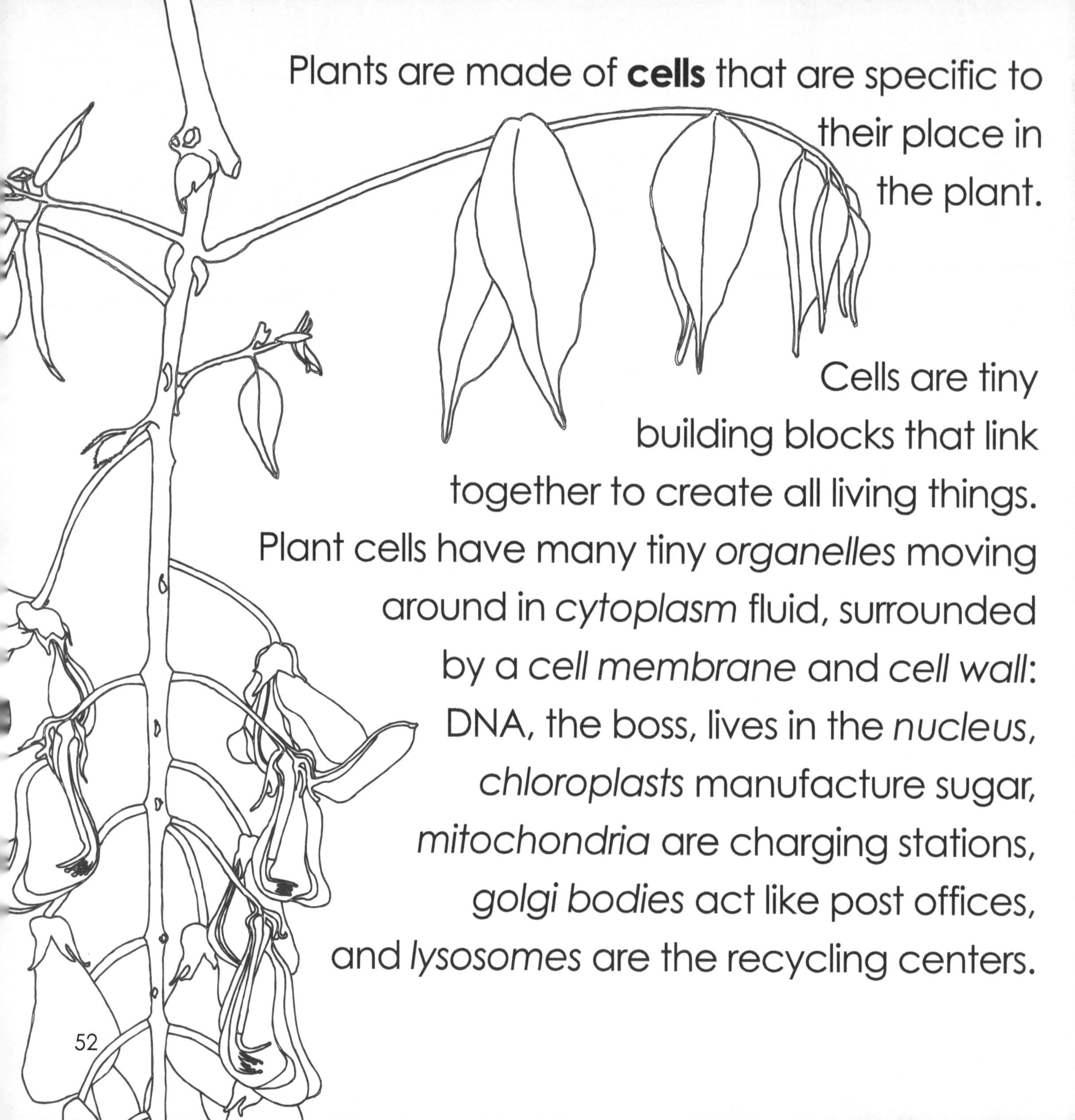

Plants are made of **cells** that are specific to their place in the plant.

Cells are tiny building blocks that link together to create all living things. Plant cells have many tiny *organelles* moving around in *cytoplasm* fluid, surrounded by a *cell membrane* and *cell wall*: DNA, the boss, lives in the *nucleus*, *chloroplasts* manufacture sugar, *mitochondria* are charging stations, *golgi bodies* act like post offices, and *lysosomes* are the recycling centers.

Many plants are used to make **medicines**, but some plants can also be **poisonous**!

All parts of the foxglove plant are poisonous. They contain a chemical that changes how human hearts beat. Doctors can use that same chemical in specific amounts to make a medicine to help people whose heart isn't beating correctly.

fun fact: Foxglove has *guard hairs* that keep out small insects.

scientific classification
genus: Digitalis
species: purpurea

foXglove

tubular flower shape
Foxglove s perfect for bumblebees.

guard hairs
prevent small
insects from entering

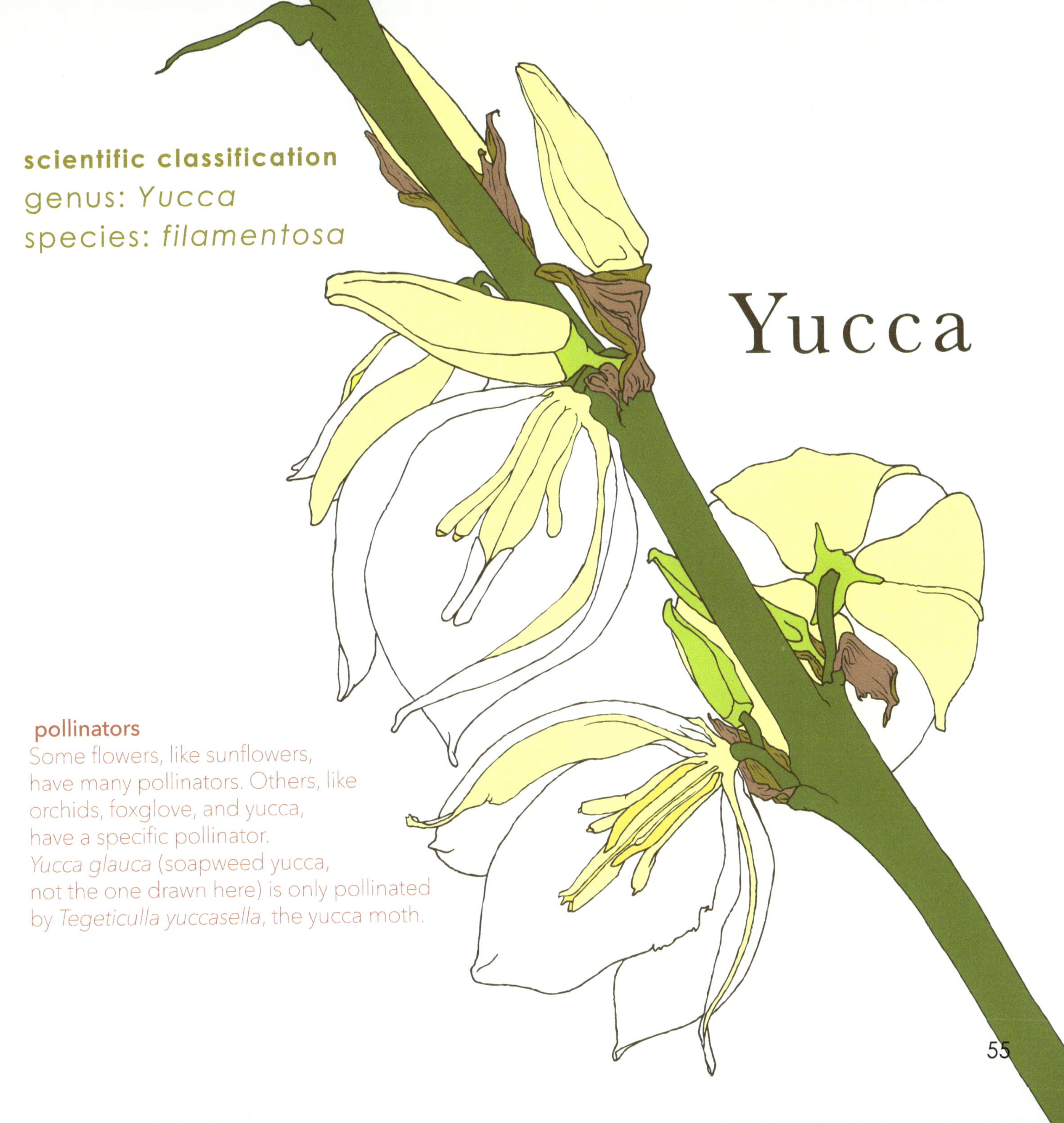

scientific classification
genus: *Yucca*
species: *filamentosa*

Yucca

pollinators
Some flowers, like sunflowers,
have many pollinators. Others, like
orchids, foxglove, and yucca,
have a specific pollinator.
Yucca glauca (soapweed yucca,
not the one drawn here) is only pollinated
by *Tegeticulla yuccasella*, the yucca moth.

55

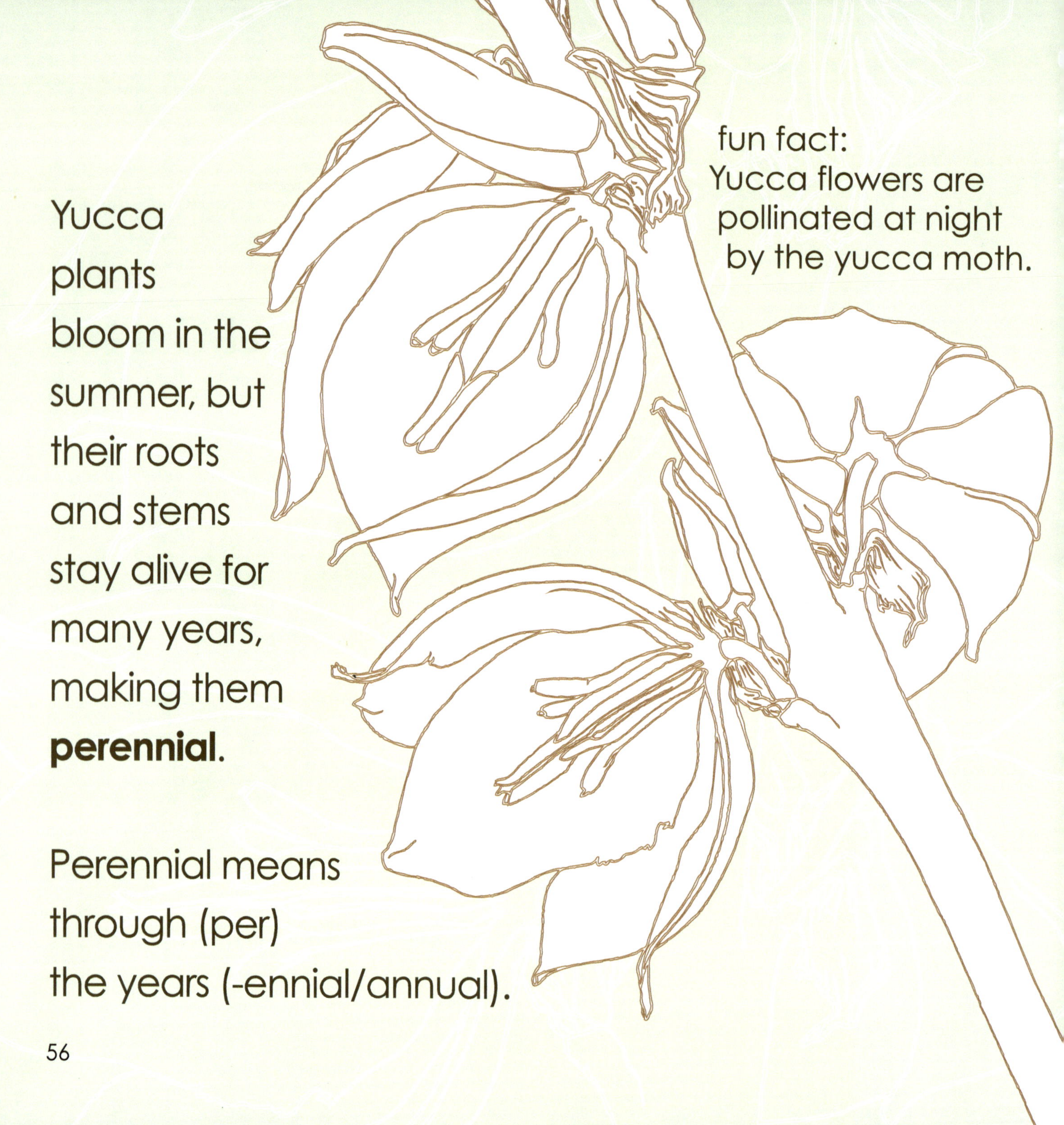

Yucca
plants
bloom in the
summer, but
their roots
and stems
stay alive for
many years,
making them
perennial.

Perennial means
through (per)
the years (-ennial/annual).

Do you see the green stem of this herbaceous zinnia plant?

Zinnias are an **annual**, so they only live for one year. The Latin word annus means year.

How can we grow zinnias next year if all of the ones we have this year die?

fun fact:
On January 16, 2016, a zinnia bloomed aboard the International Space Station as part of their work to provide food for space expeditions.

57

annuals
Many of the plants in this book are perennials, which means they live through (peri-) the years (annus). The only annuals in this book are nasturtium, sunflower, and zinnia.

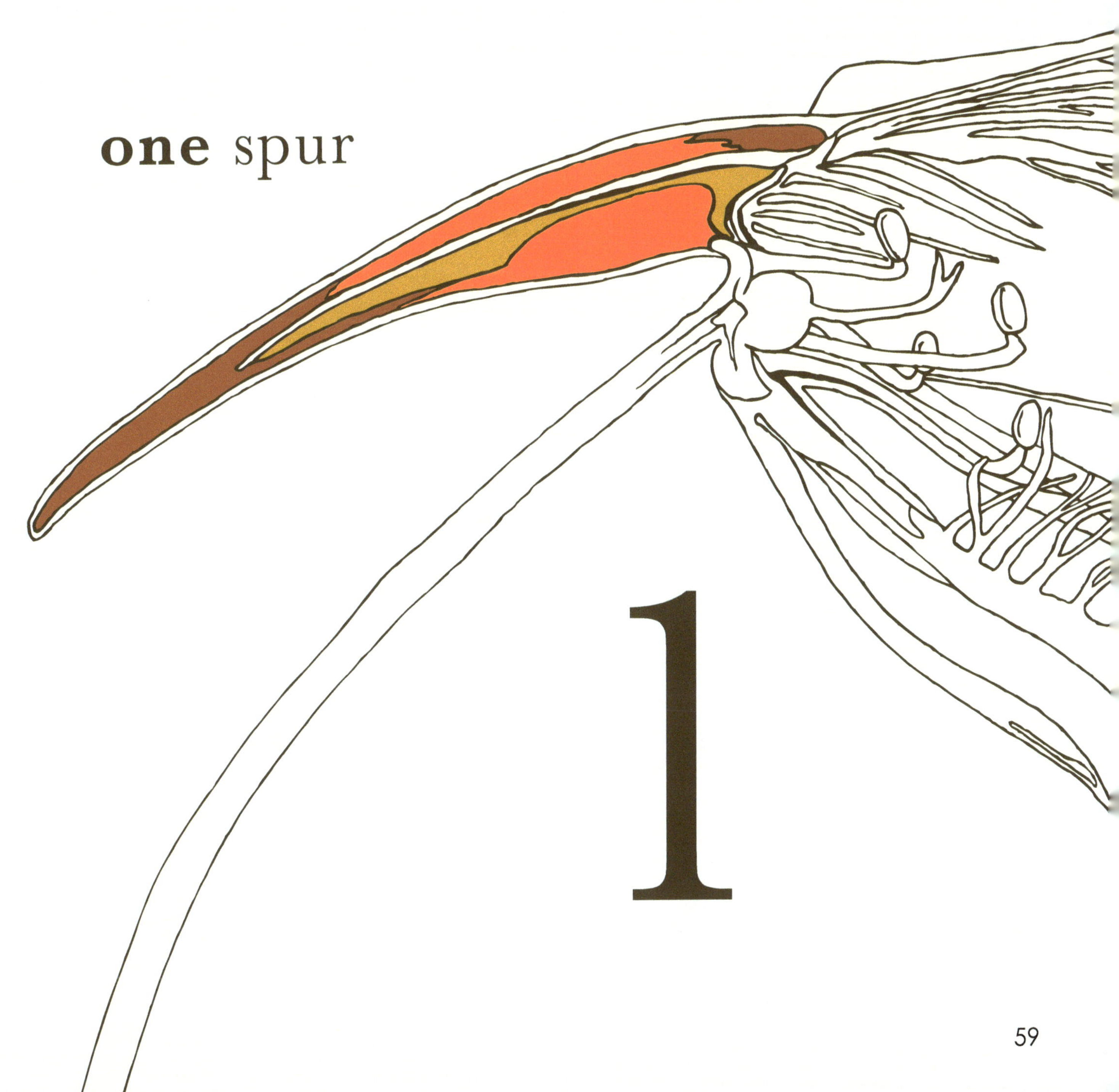

one spur
1

Nasturtiums have a long *spur* on the back of the flower where they store nectar.

Nectaries are where flowers store nectar, a sugary liquid that bees, hummingbirds, and other pollinators eat.

fun fact:
The peach flower
conceals its bright
orange nectaries
under its stamens.

bonus fun fact:
The hellebore's tube-like
nectaries have fallen off because
the flower has already been pollinated.

Many plants have thorns, spines, or prickles
to **protect** themselves.

Thorns (single) and spines (grouped) are part of the stem,
growing only from nodes, with veins like the leaves.
Prickles don't have veins and can cover the stem.

Can you see
the bumps at the base of
the thorns?

They are called nodes.——

fun challenge:
Look at the rose.
Can you figure out which
type of protection it has?
(hint: look for nodes)

2
two thorns

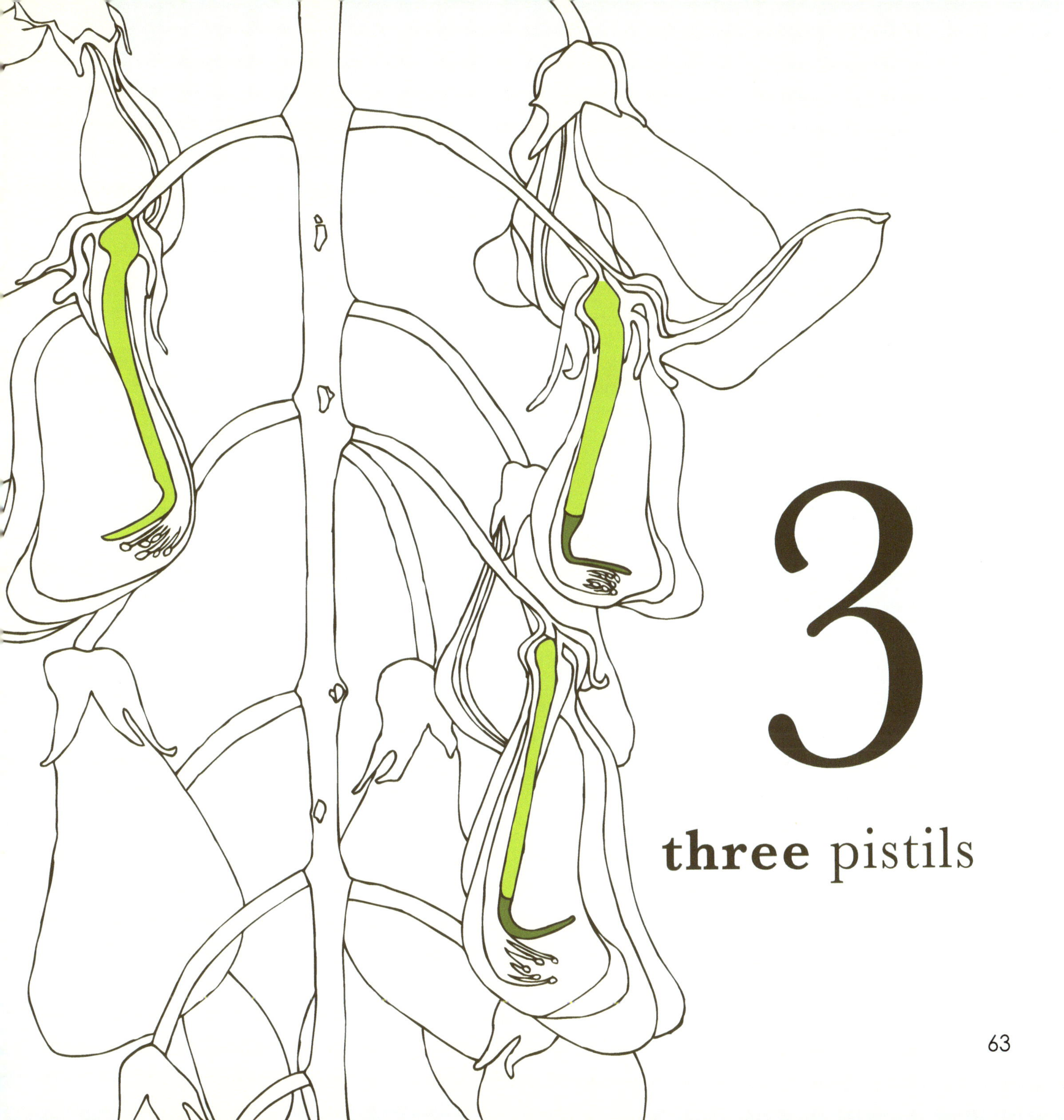

3
three pistils

The female part of the flower is called the **pistil**.

The tip of the pistil is the *stigma*. It's sticky so that pollen will attach to it.

The long, middle part of the pistil is the *style*. When pollen lands on the stigma, a tube is created to carry the pollen through the style, so the pollen "travels in style!"

The base of the pistil is the *ovary* where pollen combines with *ovules* to create seeds.

The male part of the flower is the **stamen**.
There are usually many stamens (or stamina)
and their job is to provide pollen to the pistil.

Each stamen is
made of a long
filament with an
anther on the tip
that is covered in *pollen*.

Some of the drawings in
the Aalphabet have stamens
and some don't. Why?
Stamens can fall off after they pollinate the pistil.

fun facts:
Flowers can be male, with only stamens,
Flowers can be female, with only pistils.
Flowers that have both are called "perfect" flowers.

four stamens 4

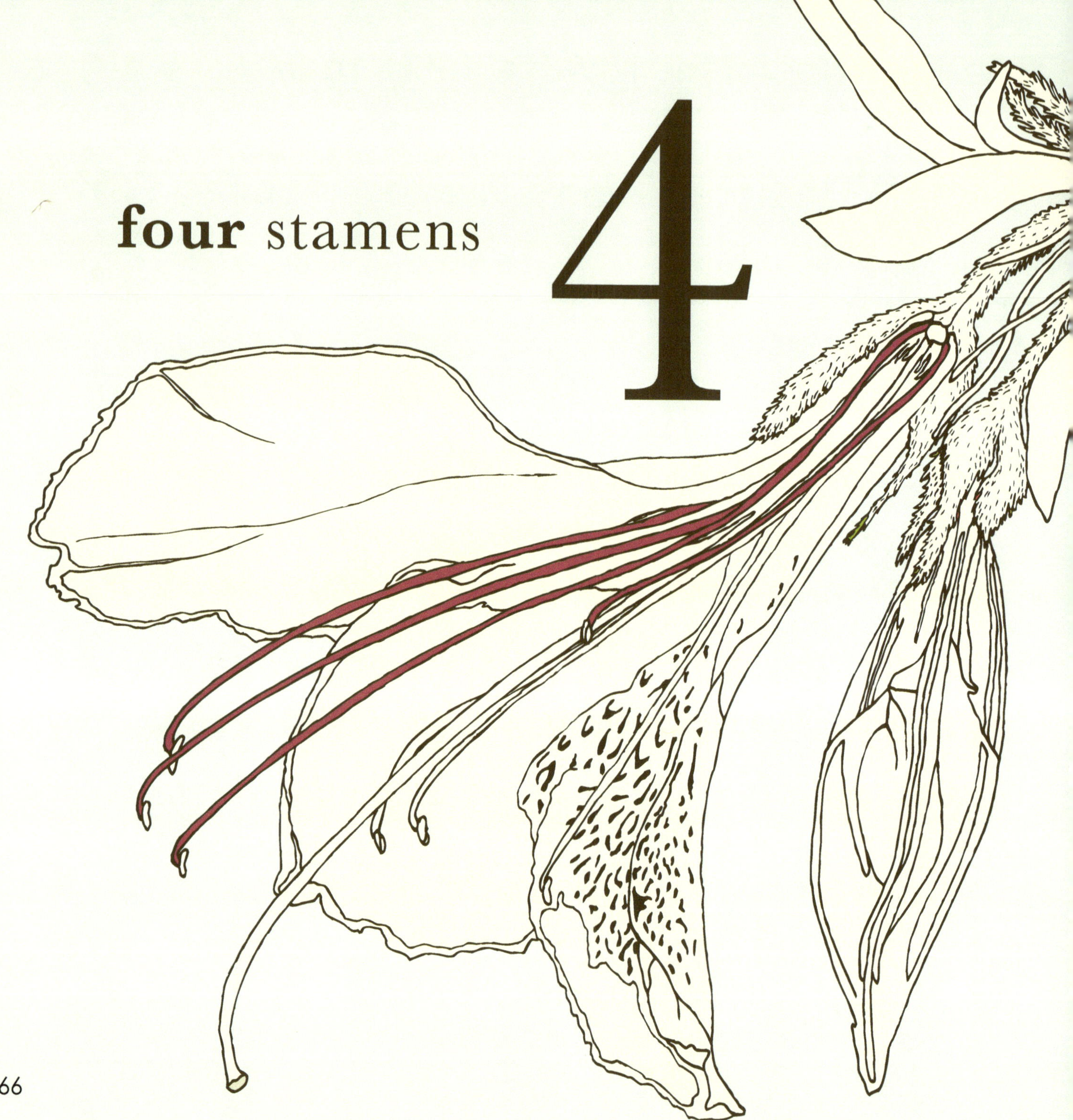

5

five tepals

Why are the "petals" of the magnolia called **tepals**?

Sepals cover petals before the flower opens, but some flowers have sepals that look like petals. When you can't tell the difference, both sepals and petals are all called tepals.

fun fact:
Magnolias are pollinated by beetles.

Christmas cacti have brightly colored **petals** surrounding the stamens and pistil. Their job is to attract pollinators so the ovules develop into seeds.

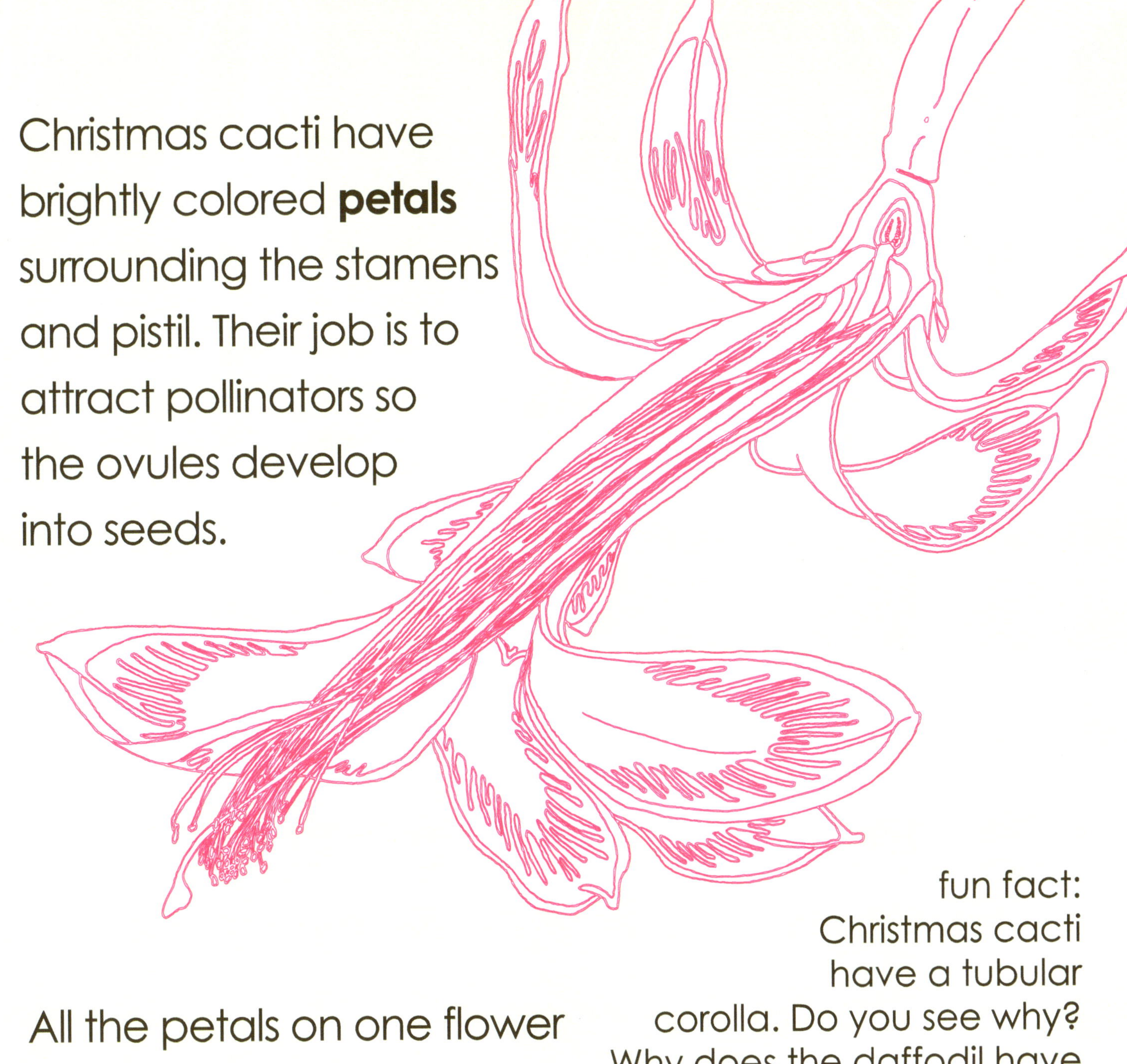

All the petals on one flower are called the *corolla*.

fun fact:
Christmas cacti have a tubular corolla. Do you see why? Why does the daffodil have a corona (crown) corolla?

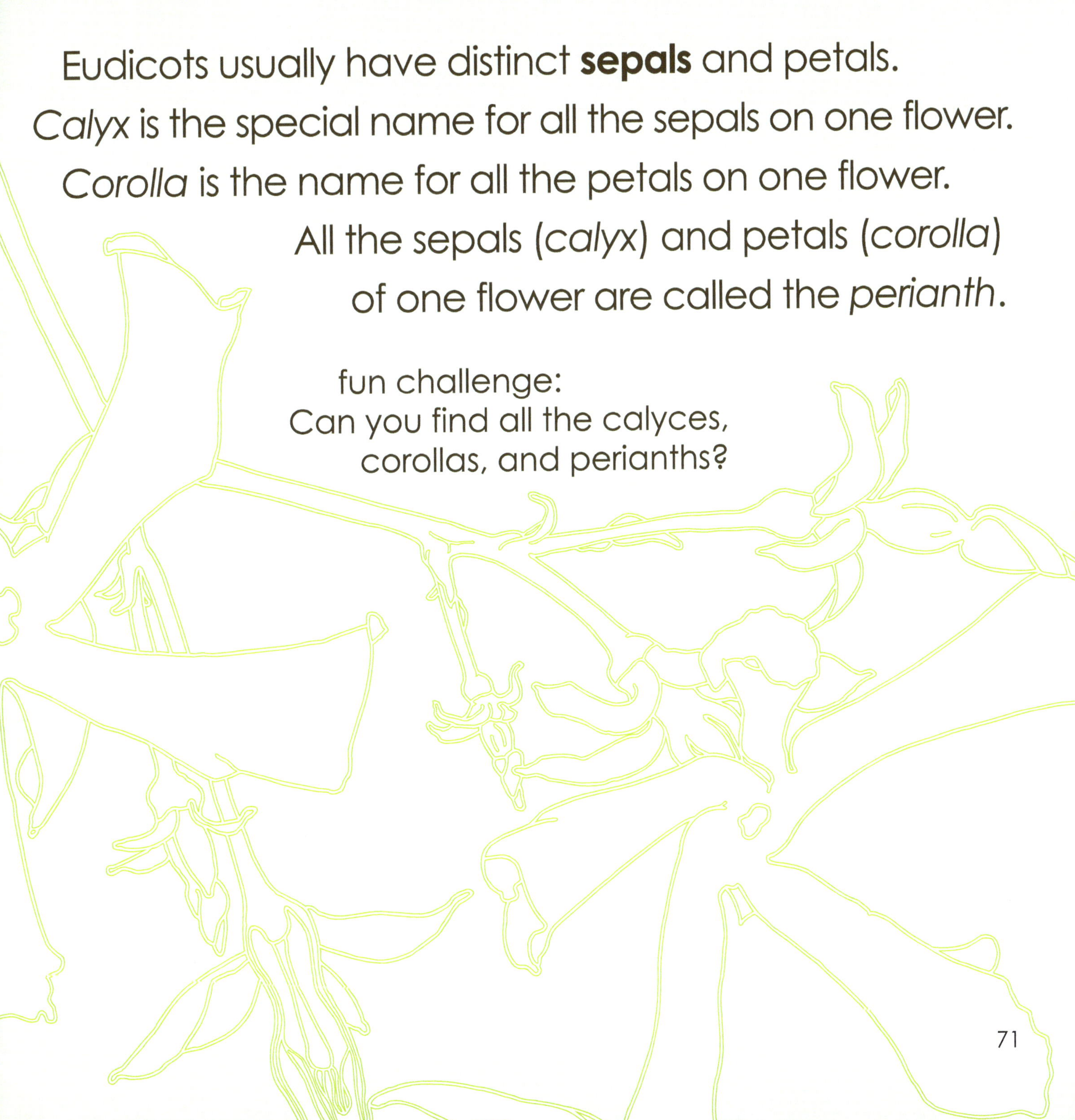

Eudicots usually have distinct **sepals** and petals.

Calyx is the special name for all the sepals on one flower.

Corolla is the name for all the petals on one flower.

All the sepals (*calyx*) and petals (*corolla*)

of one flower are called the *perianth*.

fun challenge:
Can you find all the calyces,
corollas, and perianths?

Some **bracts**, like the ones on this sunflower, protect the flower and are usually small, dry, and green. Other bracts can look like flower petals; and their job is to attract pollinators.

A bract is a type of modified leaf at the base of flowers.

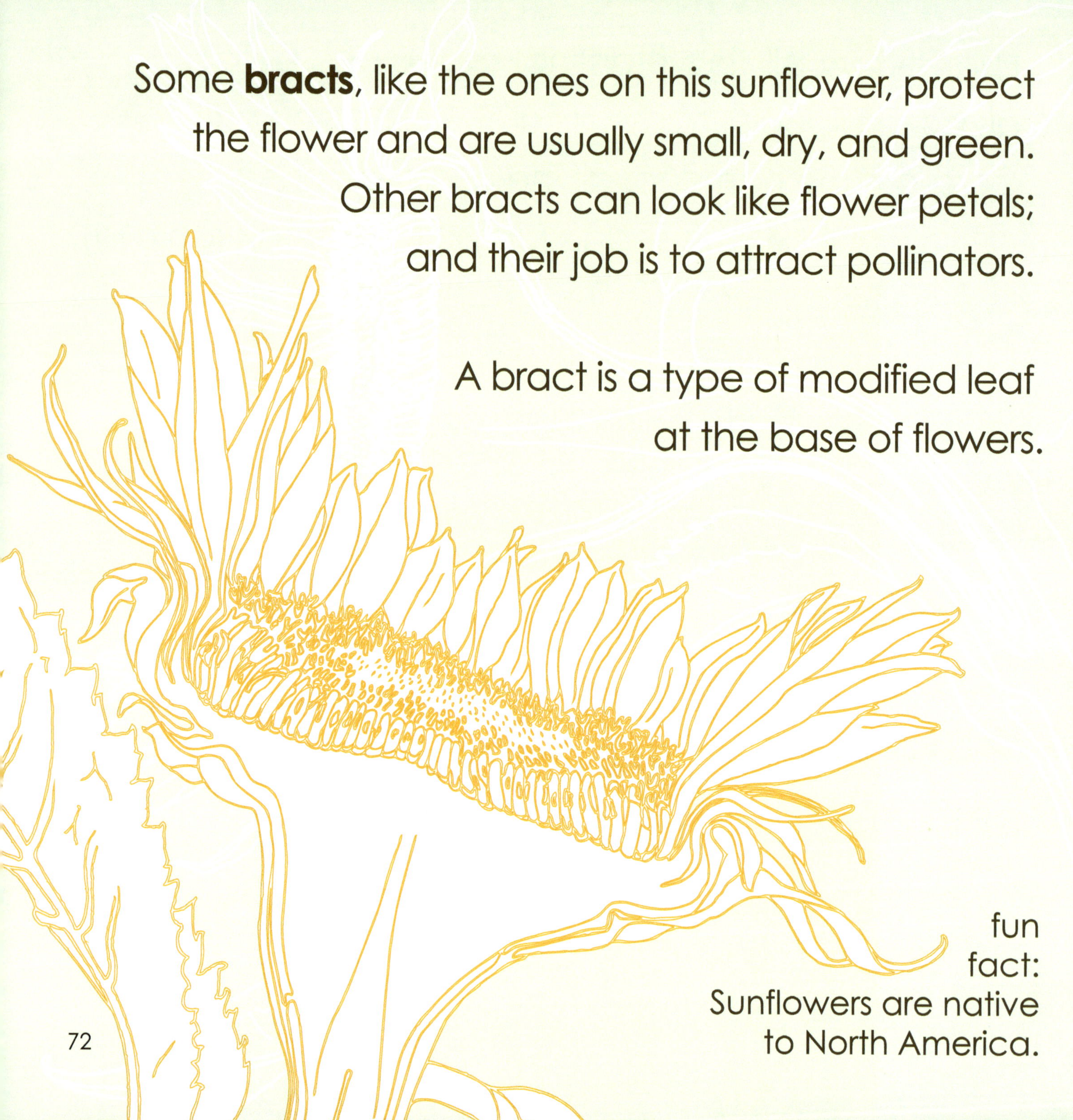

fun
fact:
Sunflowers are native
to North America.

72

Nectar is a sugary liquid that flowers produce to feed pollinators. But how do pollinators know where to find it? There are nine highlighted **nectar guides** on this toad lily tepal. The guides (dots) show pollinators, such as bees, where to find the nectar.

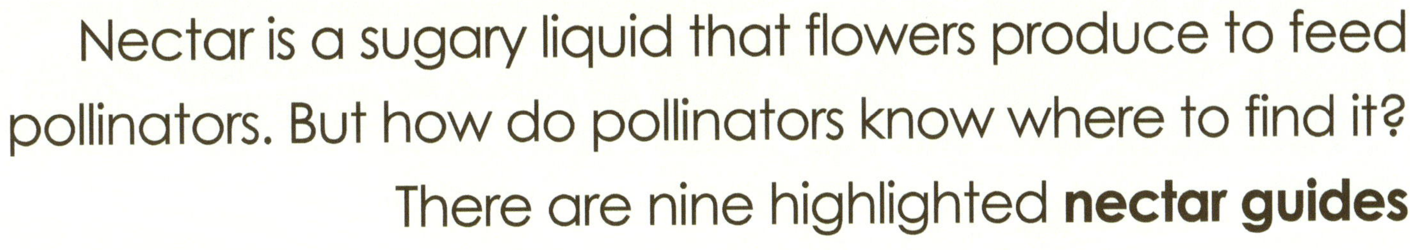

Do you see the rounded nectaries at the base of the tepals?

fun fact:
Some nectar guides are only visible to pollinators.

FLOWERING

DIVISION: Angiospermae

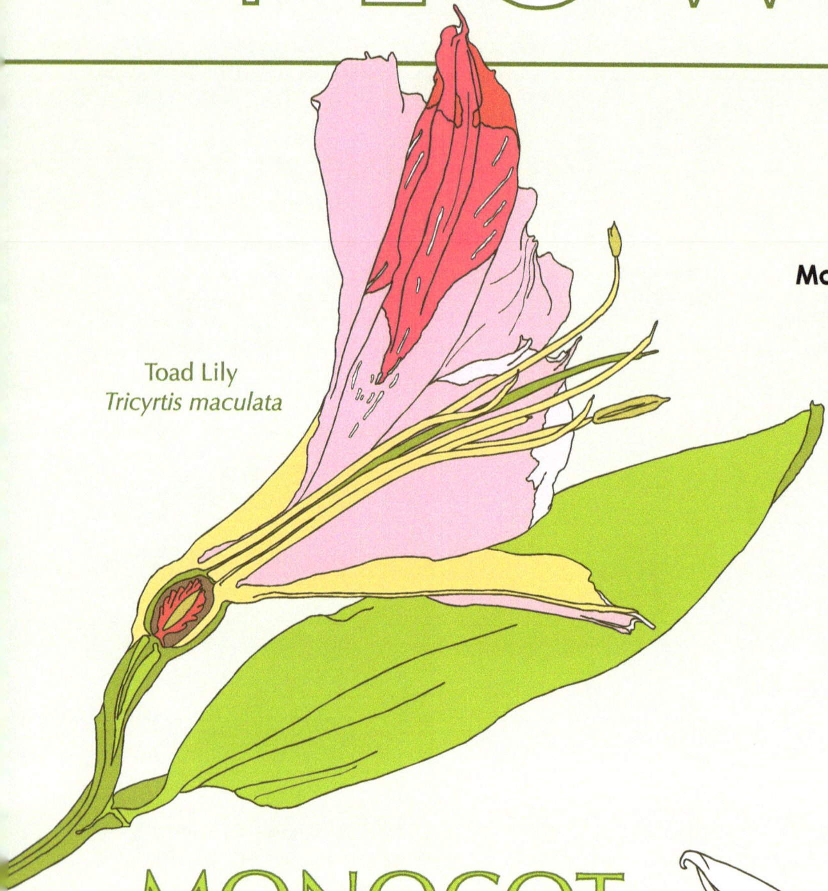

Toad Lily
Tricyrtis maculata

TEPALS, PETALS, & SEPALS

Monocotyledonae & Magnoliidae petals and sepals are so similar they are both called **tepals.** Find them in groups of **three.**

Eudicotyledonae have distinct **sepals** and **petals** that occur in groups of **four** or **five.**

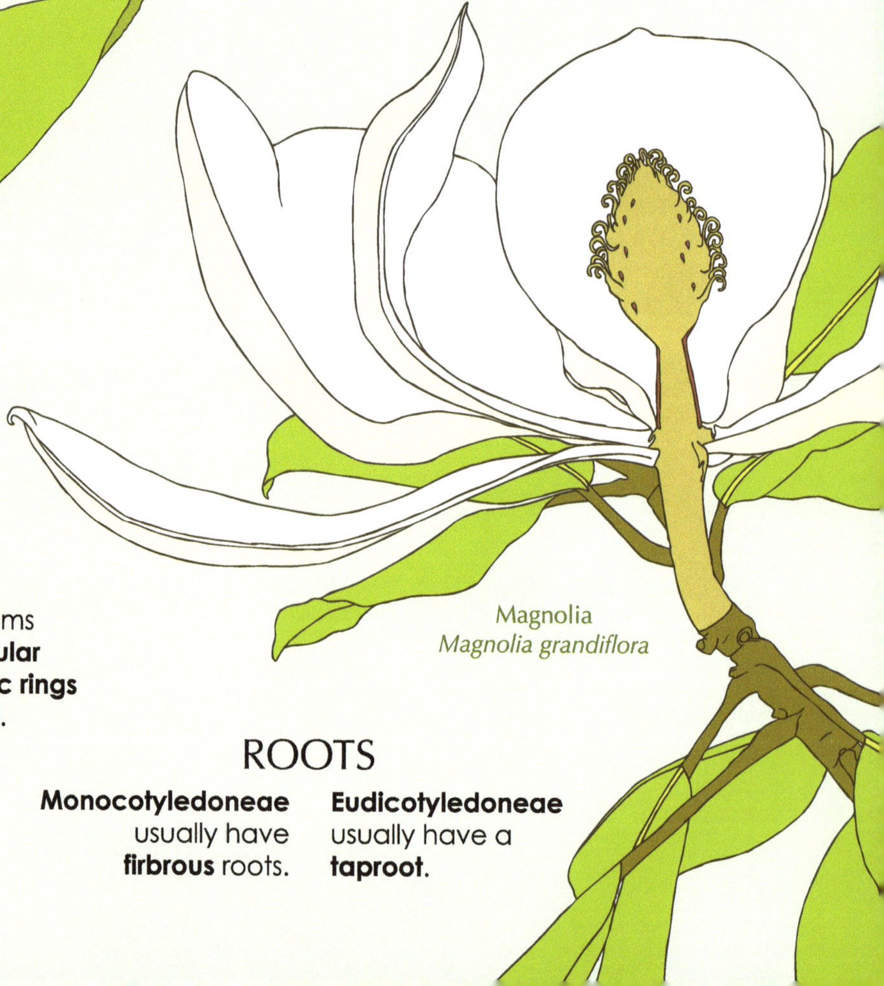

Magnolia
Magnolia grandiflora

MONOCOT

STEMS

Monocotyledonae stems usually have **dispersed vascular tissue** and are more **flexible.**

Dicotyledonae stems usually have **vascular tissue in concentric rings** making them **rigid.**

ROOTS

Monocotyledoneae usually have **firbrous** roots.

Eudicotyledoneae usually have a **taproot.**

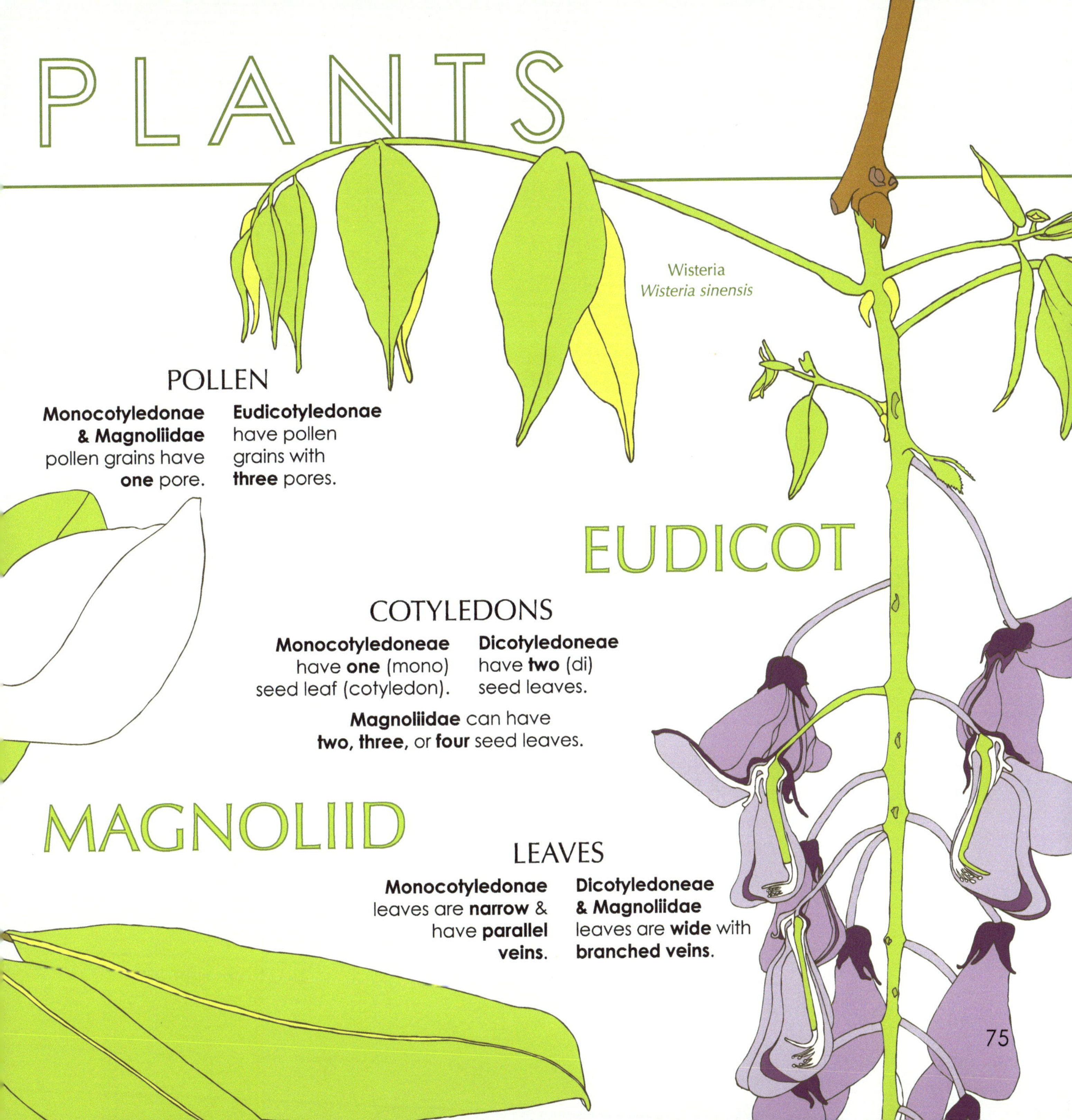

PLANTS

POLLEN

Monocotyledonae & Magnoliidae pollen grains have one pore.

Eudicotyledonae have pollen grains with three pores.

Wisteria
Wisteria sinensis

EUDICOT

COTYLEDONS

Monocotyledoneae have one (mono) seed leaf (cotyledon).

Dicotyledoneae have two (di) seed leaves.

Magnoliidae can have two, three, or four seed leaves.

MAGNOLIID

LEAVES

Monocotyledonae leaves are narrow & have parallel veins.

Dicotyledoneae & Magnoliidae leaves are wide with branched veins.

75

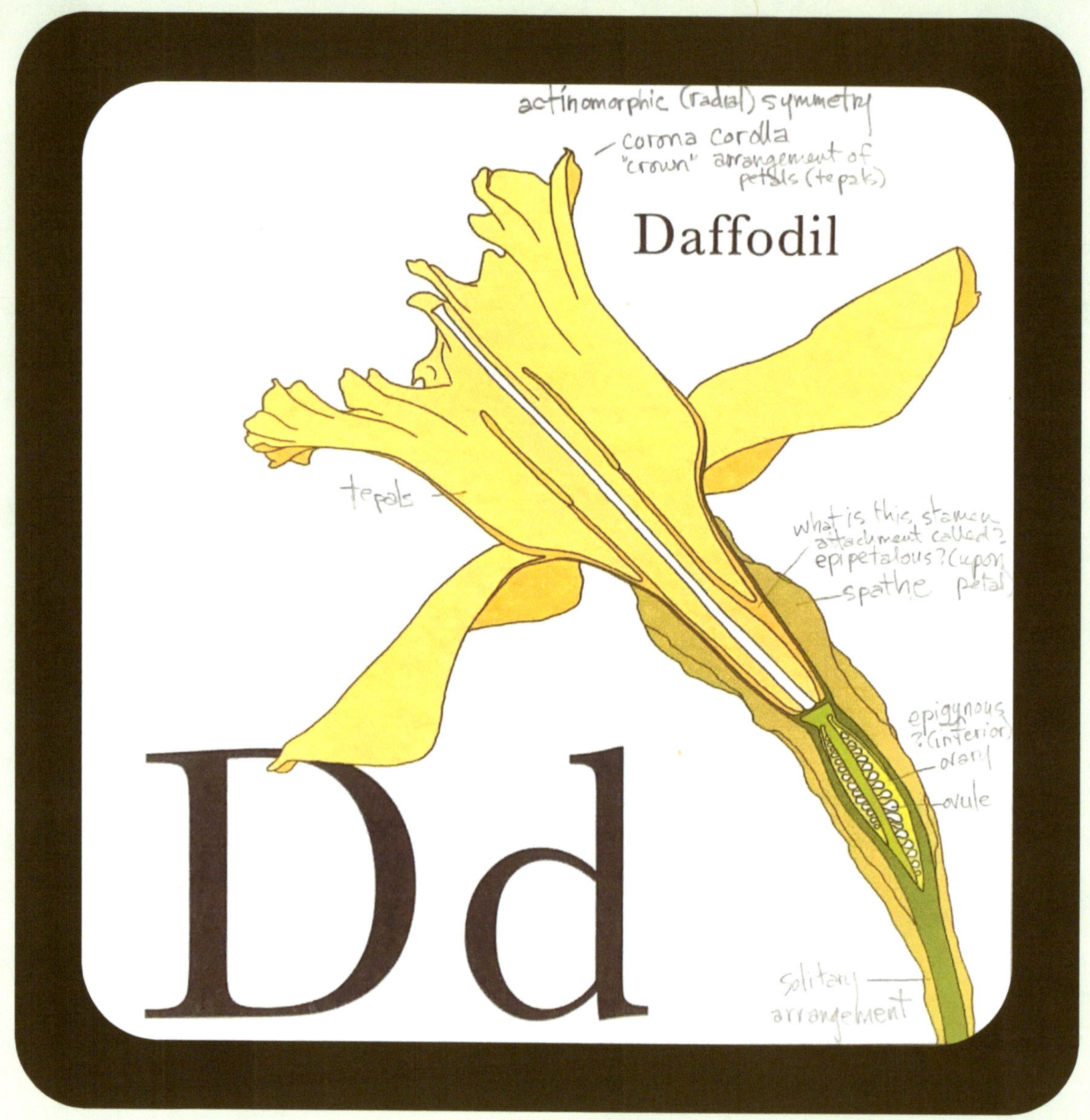

front of *Eileen's Dd Art Card*
drawings and notes from spring 2020, uncoated cardstock 6" x 6"

Facsimiles are for insight and inspiration only. These represent the actual inquiry process. Notes are not fully researched and may be unreliable.

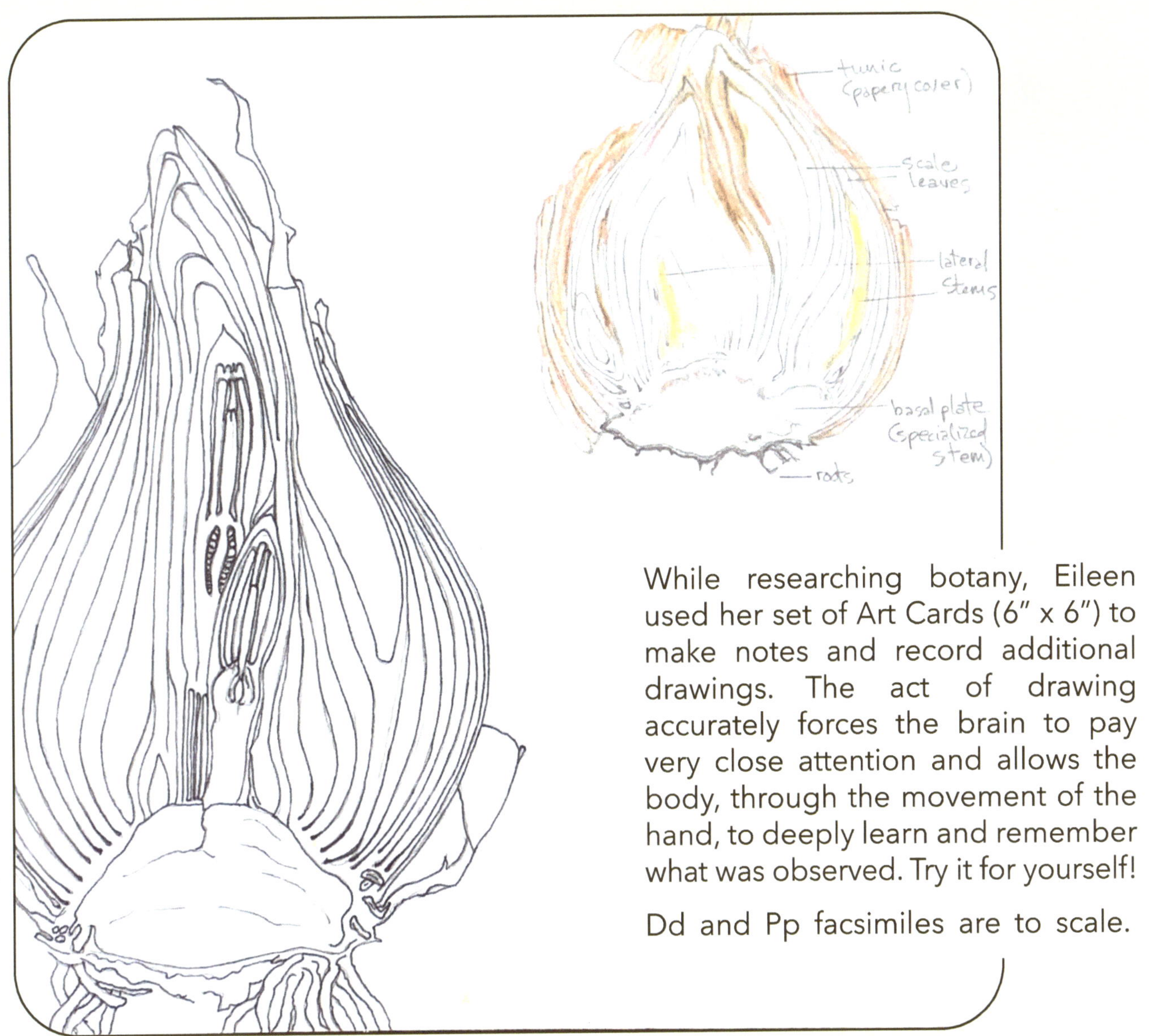

While researching botany, Eileen used her set of Art Cards (6" x 6") to make notes and record additional drawings. The act of drawing accurately forces the brain to pay very close attention and allows the body, through the movement of the hand, to deeply learn and remember what was observed. Try it for yourself!

Dd and Pp facsimiles are to scale.

These bulbs were cut for the Daffodil Dissection Video. What a surprise to find tiny flowers hidden within! After discovering and documenting their hidden treasures, Eileen planted the larger halves of the cut bulbs in the front garden bed. Despite the damage, they continued to grow and bloom.

Plants are capable of surviving many types of trauma, often prioritizing reproduction when they are dying. Christmas cacti will send out adventitious roots at every section of their stem when stressed. Magnolia will grow new trees from roots. Jade will flower when the roots are severely compacted.

front of *Eileen's Pp Art Card*
drawings and notes from spring 2020, uncoated cardstock 6" x 6"

Facsimiles are for insight and inspiration only. These represent the actual inquiry process. Notes are not fully researched and may be unreliable.

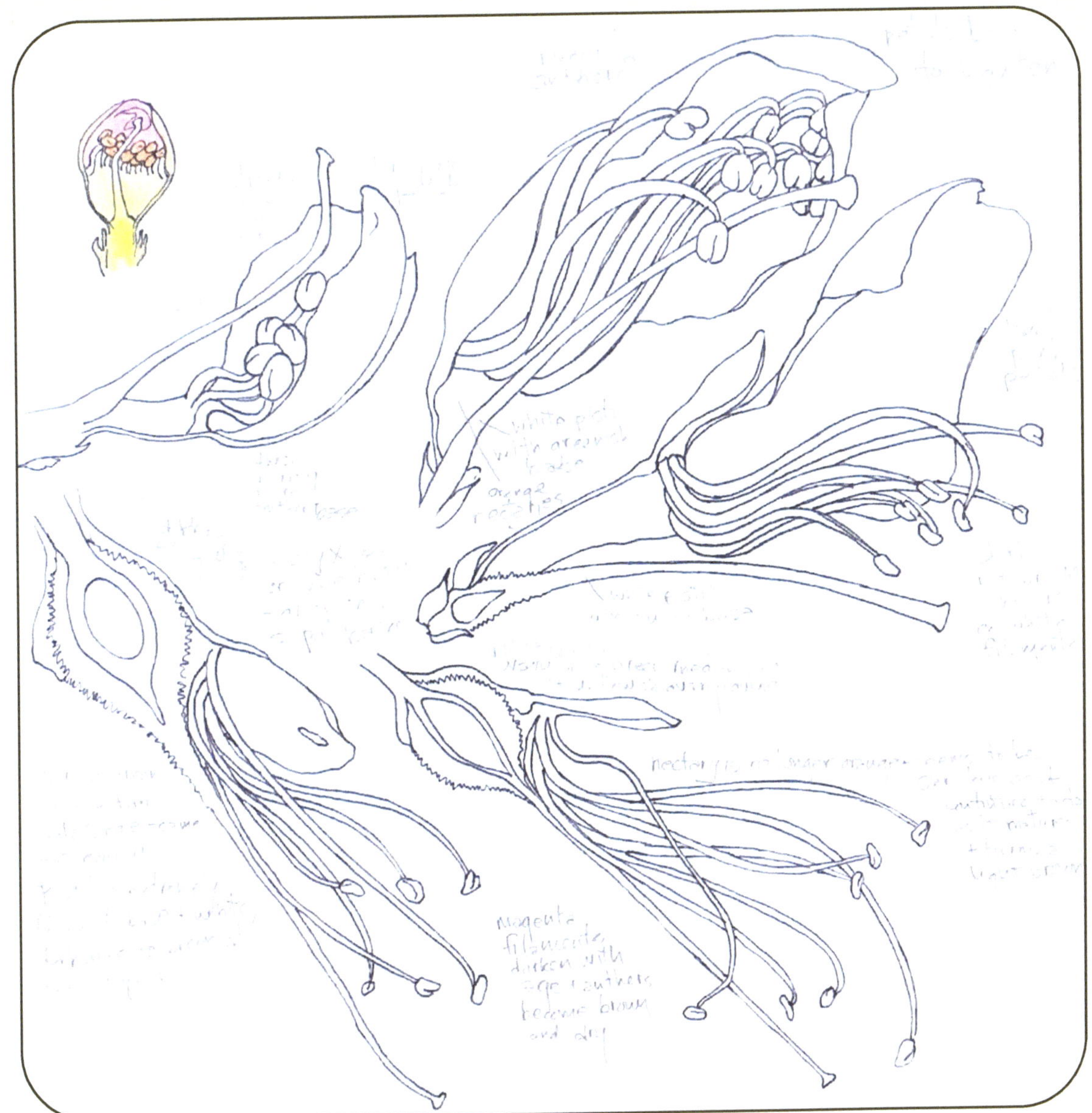

Tips for note taking:
* Record details, even then ones that may seem obvious now. In the future, this will be the only information available.
* Write down questions that come to mind. Not only does this provide a focus and establish connections, it is a record of how much was known or unknown. Humans are always discovering, so recognize that all information is incomplete and possibly inaccurate.

GLOSSARY

angiospermae - scientific name for flowering plants. *Angio* means *covered* and *sperm* means *seed*. Angiosperms make covered seeds with their fowers.

annuals - plants that complete their life cycle and die in one year.
annuals: nasturtium, sunflower, and zinnia

binomial nomenclature - the naming system that Carl Linneaus created to identify plants. It is used to identify the genus and species of all living things. *Bi* means *two* and *nomial* means *name*, so each classification has two names. The first word (*name*) is the genus and the second word (*name*) is the species. The genus is capitalized and species is lower case. Both words are italicized. ie. *Homo sapiens* (human) and *Dicentra formosa* (bleeding heart)

bract - modified leaf at the flower base. Bracts can be showy (the red petal-like parts of the poinsettia), dry and protective (sunflower), or papery (daffodil).

bulb - underground stems attached to special scale leaves that store food. Bulbs produce flowers, then die back and reemerge the following year. Bulbs are usually herbaceous perennials.

calyx - all the sepals on flower

cell - smallest unit of life, containing many organelles that depend on each other to gather, store, or use energy, and do work according to the directions the cell contains

clade - group of plants that have similar characteristics

corolla - all the petals on a flower

deciduous - plants that lose their leaves in winter and regrow them in the spring
woody deciduous: forsythia, kudzu, peach, rose, wisteria

evergreen - plants that have leaves all year long
woody evergreen: azalea, gardenia, jasmine, magnolia, quince (deciduous in cold climates), yucca

florets - tiny flowers that make up a single flower. **Disc florets** can produce seeds and are in the center of the flower. **Ray florets** are much larger and usually have fused petals that look like individual petals around the edge of the flower. Sunflowers and echinacea flowers have florets.

flower - seed-producing part of a plant that has petals and sepals (or tepals), stamen/s, and pistil/s. Tiny flowers in a capitulum inflorescence are florets. Before the flower opens, it is called a bud.

fruit - final stage of the ovary, containing the seeds of the plant.
NOTE: Some vegetative plant parts are called "fruits." The juicy flesh of a strawberry is not part of the ovary, so it is not the fruit. Strawberries have dry fruits; the covering of the seeds are hard and dry.
ALSO NOTE: Some fruits are called "vegetables." Squash and cucumbers are fruits because they are the fully developed ovaries of the plant and contain the seeds. All seeds are inside the fruit.

inflorescence - how flowers are arranged on a stem. A solitary inflorescence has a single flower on a stem (daffodils). A capitulum inflorescence has many florets side-by-side (sunflowers & echinacea).

leaves - flat structures extending from the stem that make sugar for the plant. Leaves are classified by venation (how veins are arranged), margin (edge), and shape (silhouette).

life cycle - *cycle* means *circle*, so there is no beginning or end. For flowering plants, the process is: Seeds sprout and turn into seedlings. Seedlings become plants. Plants produce flowers that are pollinated, creating seeds. Not all plants produce seeds; some use a different process with spores.

nectar - sweet substance that flowers use to feed animals in exchange for pollination, often stored in a **nectary**. Plants often use special coloration, called **nectar guides**, to attract pollinators and direct them to the nectar.

ovary - in flowering plants, the ovary becomes the fruit. Tiny **ovules** inside the ovary become seeds.

perianth - *peri* means *around* and *anth* is short for *anthos*, the Greek word for flower. The perianth consists of the calyx (all the sepals of one flower) and corolla (all the petals of one flower).

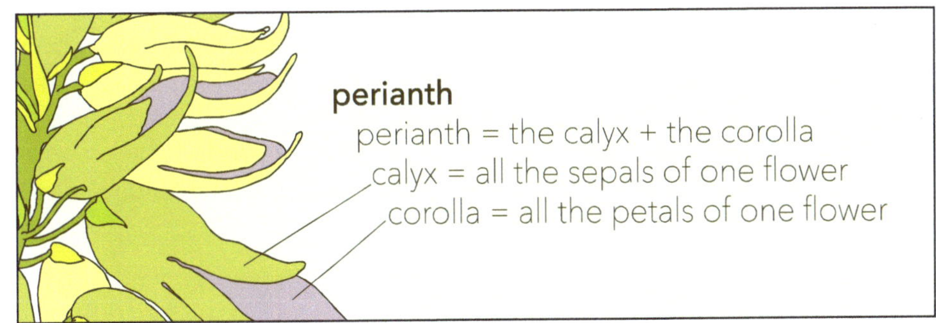

perennial - plants that come back year after year; herbaceous or woody.
woody perennial: azalea, forsythia, gardenia, jasmine, kudzu, magnolia, peach, quince, rose, wisteria,
herbaceous perennial: bleeding heart, daffodil, echinacea, hellebore, iris, lily, orchid, toad lily, Ulster Mary, violet, foxglove

petal - showy flower parts surrounding the stamens that attract pollinators; petals are found in groups of 4 or 5 on eudicots.

photosynthesis - a complex process in which special cell parts–chloroplasts–gather and store energy from the sun. Other plant parts can transport the energy packets to the entire plant. The basic formula is sunlight + carbon dioxide + water = sugar.
FUN FACT: Plants break water into hydrogen and oxygen, then makes new water by combining hydrogen and oxygen back together!

pistil - female part of the flower consisting the stigma (sticky tip), style (long connector), and the ovary (where seeds develop).

pith - soft, spongy tissue that stores nutrients.

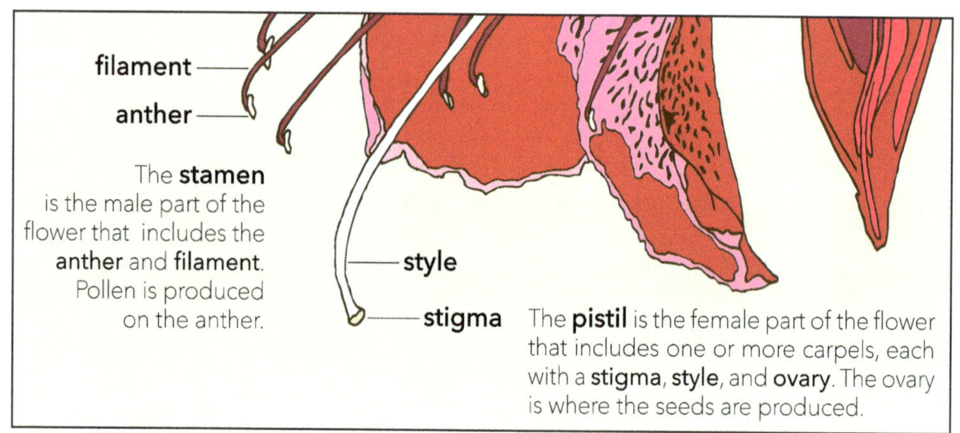

plant - multi-celled living orgaism that uses photosynthesis to create packets of energy (sugar)

pollen - tiny grains that ovules need to make seeds. When pollen lands on the sticky stigma, the pistil makes a tube in its style to bring the pollen down to an ovule and create a seed.

pollination - transfer of pollen from anter to the stigma. Plants can be pollinated from the placement of the stamen and pistil (bleeding hearts), by movement of wind (grasses), or by pollinators.

pollinators - animals that transfer pollen from the anther to the stigma of the pistil, allowing the flower to produce seeds. Some flowers cater to particular pollinators with specific defenses (ie. guard hairs) and shapes (ie. tubular).

prickle - sharp points that emerge all along the stem but are not connected to the vascular system. NOTE: Roses have prickles, not thorns.

reproduction - creation of new living organisms using cells and genetic material from one or two existing organisms.

reproduction, sexual - generation of new plants from seeds that are created by the sexual parts (stamen, pistil of an existing plant or plants.

reproduction, vegetative - generation of new plants from the vegetative material (roots, stems, leaves) of existing plants

rhizome - underground or partially underground segmented stems that can be used to self-propagate the plant. Iris often have partially underground stems.

roots - plants parts, usually at the base of the plant, that collect nutrients and usually keep the plant securely in one spot. Roots can be underground, float in water (ie. water lilies), and grab on to other plants (ie. orchids). Orchid roots can make food for the plant (photosynthesis) like leaves.

seeds - plant embryo and its nutrition surrounded by a protective seed coat.

sepal - layer of petal-like or leaf-like parts that protect the petals as they develop. The group of all sepals on a flower is called the calyx.

stamen - male part of the flower that includes pollen on the anther (where pollen forms) and a filament (long, thin attachment to the flower).

stem - central part of the plant that provides structure; can be woody or herbaceous

tepals - petals and sepals that are indistinguishable. Most monocots (lily, orchid) and some magnoliids (magnolia) have tepals.

thorns - single sharp points that emerge at nodes along the stem. Quince has thorns.

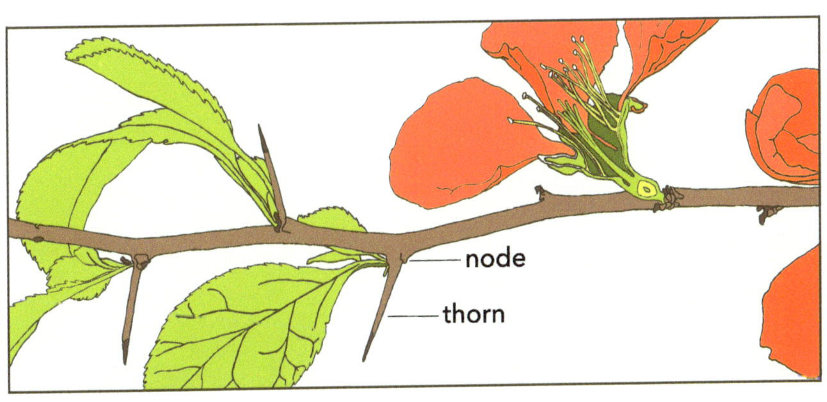

BOTANY BOOKS TO TREASURE

A Seed is Sleepy, Diana Hutts Aston & Sylvia Long

Flora: Inside the Secret World of Plants, DK

Gregor Mendel: The Friar Who Grew Peas, Cheryl Bardoe

Grow Your Soil, Diane Messler

Karl, Get Out of the Garden! Anita Sanchez

Natural History, DK Smithsonian

Ultimate Visual Dictionary, DK

Eileen Chevalier's TYPEWRITER Aalphabet BOTANY Curriculum

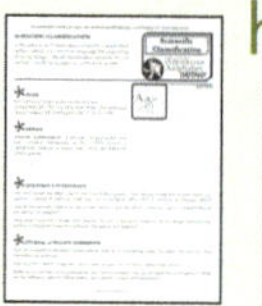
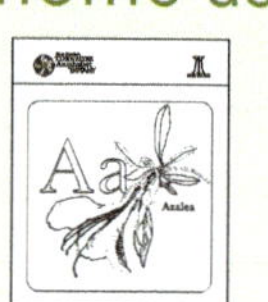

Teacher's Guide

Curriculum overview, lesson guides, coloring pages, glossary, & more.
8.5 x 11 in. 80 pgs. paperback.
ISBN 978-1-952704-13-0

Flashcards

A-Z & 0-9 illustrations & facts
4.8 x 4.8 in. 74 pgs. looseleaf.
ISBN 978-1-952704-01-7

Botany Workbook

36 botany lesson prompts on graph paper.
6 x 9 in. 176 pgs. paperback.
ISBN 978-1-952704-15-4

Color +Label

Black outline versions of the A-Z & 0-9 illustrations with labels, definitions, botany facts, and glossary/index.
8.5 x 8.5 in. 38 pgs. paperback.
ISBN 978-1-952704-14-7